*The School  of Chemical Engineering at Cornell*

CORNELL ENGINEERING HISTORIES
Volume 1

*Fred H. "Dusty" Rhodes, founder and first director of the School, wearing his "old school tie." (See page 17.)*

*To the memory of my
former teacher, employer,
mentor, critic, and friend,
Fred H. "Dusty" Rhodes*

# The School of Chemical Engineering at Cornell

## A History of the First Fifty Years

*Julian C. Smith*

College of Engineering, Cornell University
Ithaca, New York

1988

© 1988 by the College of Engineering, Cornell University
Ithaca, New York 14853–2201

Printed in the United States of America

Library of Congress Catalog Card Number: 88-70502.

ISBN 0-918531–02–0

# Contents

# Foreword

This is a history of the School of Chemical Engineering at Cornell University, from the time of the School's founding and establishment as part of the College of Engineering in 1937–38 to the present, fifty years later. For all but a few of those years, I have been directly associated with the School as student, faculty member, director, and professor emeritus. I have known all the faculty members the School has had, and most of its alumni. A wealth of material has been available to choose from for this history; its very richness made it difficult to know what to include, what to omit. I have tried to give a reasonably balanced, accurate account of the way things were, but undoubtedly some people and some events—I hope not many—have not been given proper consideration. If there are errors or glaring omissions, I would be glad to hear of them.

The origins of chemical engineering at Cornell were described at length in "The Development of the Department of Chemistry and of the School of Chemical Engineering at Cornell," by E. M. Chamot and F. H. Rhodes, written in 1957 but never published. The first five chapters of my history are largely based on that document. Other sources include annual reports to the president of the University; faculty registers and staff directories available in the John M. Olin Library at Cornell; *A History of Cornell* by Morris Bishop (Cornell University Press, 1961); the *Olin Hall News,* 1946 to 1987; faculty minutes, 1963 to 1978; annual reports and other records of the School of Chemical Engineering; miscellaneous archival files; and, of course, the memory banks of many professors and alumni. The pictures came from the Cornell Department of Manuscripts and Archives, the College of Engineering, the School of Chemical Engineering, and from my own collection and that of Professor Raymond S. Thorpe.

To all who contributed to this undertaking, many many thanks. I especially appreciate the careful reviews of the manuscript by Nancy Morris, Professor George F. Scheele, and Director Keith E. Gubbins. Special thanks go also to Gould P. Colman of the Department of Manuscripts and Archives for his suggestions and assistance; to Professor Thorpe for the loan of his photographs and faculty files; and to Gladys McConkey and David Price of the

Office of Publications of the College of Engineering, who did the final editing and the design. The cooperation and help I received made the writing of this history a pleasure instead of a chore. I hope these chapters convey a sense of the unusual origins and distinctive qualities of the School of Chemical Engineering at Cornell, for in many ways it is unique. Finally, I can say just one more thing about the School: it has been a great place to spend the last fifty years.

Julian C. Smith
Ithaca, New York
January 1988

# 1. The Beginnings

Although chemical engineering as a separate discipline in the College of Engineering at Cornell began in 1938, its roots go back to the very early days of the University.

In 1870, two years after instruction commenced, C. H. Wing was appointed Professor of Chemistry Applied to Manufacture. In 1891–92 an option in chemical engineering was offered to students in mechanical engineering, but it aroused little interest. In 1901 Robert Henry Thurston, director of the Sibley College of Mechanical Engineering and the Mechanic Arts, advocated without success the formation of a curriculum in chemical engineering. Various courses called Industrial Chemistry were given from time to time in the Department of Chemistry. Each year from 1901 to 1905, Dr. H. R. Carveth gave a one-term course described as "an elementary introduction to chemical engineering." Apparently, however, this was a purely descriptive discussion of industrial equipment and practice in the limited chemical industry of the day. This course offering ended with Dr. Carveth's resignation from Cornell in 1905. Oddly, these early courses were available only to engineering students and not to chemistry majors.

In early years there was a professional program in chemistry leading to the degree Bachelor of Science in Chemistry, but it was discontinued in 1890. For the next twenty years the degree for chemistry majors was Bachelor of Arts. By 1911, however, many professors in the humanistic fields of the College of Arts and Sciences, of which Chemistry was a part, had come to feel that much of the work in chemistry was technological rather than cultural or truly scientific, and that credit toward a Bachelor of Arts degree should not be granted for such work. In the ensuing squabble the faculty of Chemistry discussed such things as reviving a professional degree in chemistry, the formation of a separate College of Chemistry, and even the possibility of establishing a degree program in chemical engineering.

The first curriculum in chemical engineering anywhere in the world was established at MIT in 1888, and by 1911 a few other American universities had instituted such programs. The American Institute of Chemical Engineers (AIChE) had been founded in 1908. Even so, Professor Louis M. "King" Dennis, chairman of the Cornell Department of Chemistry, stoutly maintained that there was no such thing as a chemical engineer. There was also apprehension in the faculty of Chemistry that the technical aspects of chemical engineering would overshadow the scientific aspects and diminish the importance of teaching and research in chemistry. After much debate, the faculty of Arts and Sciences approved a separate technical curriculum leading to the



*[The] chairman of the Cornell Department of Chemistry. . . stoutly maintained that there was no such thing as a chemical engineer.*

*Above: Baker Laboratory of Chemistry. The first instruction in chemical engineering was given here.*
*Below: The entrance to the Baker parking lot (1941). The Chemical Engineering office was just to the right of the archway.*

special degree of Bachelor of Chemistry, with responsibility for the program given to the Department of Chemistry. This unusual professional program in the College of Arts and Sciences led, in a rather devious way, to the establishment of Chemical Engineering at Cornell.

In 1916–17 Dr. Fred H. "Dusty" Rhodes, a newly appointed instructor in Chemistry, offered a course called Industrial Chemistry as a substitute for a chemistry course that was to have been given in Morse Hall but which had been cancelled because of the fire that destroyed the building. His qualifications were that he had done a little work for Anaconda Copper and had written a book on the subject. According to Dr. Rhodes, it was "a pretty good course." But the next year he left Cornell for the Chemical Division of the Barrett Company, where he quickly rose to the position of director of research. There he became, through hard-earned experience, a chemical engineer. He returned to Cornell in 1920 as Professor of Industrial Chemistry, then a required subject for the Bachelor of Chemistry degree. From the date of his reappointment his lectures, though officially on industrial chemistry, were in fact lectures in chemical engineering.

Serious discussion of a curriculum in chemical engineering was resumed in 1928. By that time the profession was so firmly established and the demand for chemical engineers in industry was so great that for Cornell to maintain its position as a leading university in engineering it had to offer a degree in chemical engineering. Its natural home was the Department of Chemistry, but a number of chemistry faculty members opposed it, apprehensive that it would entice many students away from chemistry. There were no funds in the College of Engineering to support such a venture anyway, so when Dr. Rhodes suggested a five-year program— four years in Chemistry and a fifth in Engineering— the suggestion was received with enthusiasm. The chemists were sure that a five-year program would not succeed (there were no five-year engineering programs at that time); the engineers were also confident that it would fail, and that even if it didn't, the College of Engineering would have to deal with only a few new students, and those few at the graduate level.

Thus in 1930 a curriculum in chemical engineering was approved, with four years in Arts and Sciences leading to the Bachelor of Chemistry degree, followed by one year in Engineering leading to the degree of Chemical Engineer. Actually, two curricula were approved: one for students who wanted only the B.Chem. degree, and the other for would-be chemical engineers. Admission to the fifth year depended on completion of all prerequisite courses and maintenance of a satisfactory average. In other branches of engineering the minimum acceptable average was 70 percent, but for chemical engineers Dr. Rhodes interpreted "satisfactory" to mean 75 percent. He alone determined whether a given student could or could not enter the fifth year. (For some the trauma of denial was very great: one would-be chemical engineer who was not permitted to continue wrote frequent letters to administrators of the School, the College, and the University for the next forty-five years, bitterly complaining of the "unjustifiable, arbitrary and capricious" treatment he had received.)

An item from Dr. Rhodes' history of the early days is revealing of his methods. "Control over the new curriculum and of the students in the fifth year was vested in a committee composed of Professors Papish and Johnson from Chemistry, and Director Diederichs of Mechanical Engineering, with Professor Rhodes as chairman. Since this gave a preponderance of voting power to the representatives from the Chemistry Department, Professor Rhodes suggested that Professor A. C. Davis, from Mechanical Engineering, be added. This gave equal representation to the two divisions. The voting on most questions of policy found two members on each side; the chairman was called upon to cast the deciding vote. Since it didn't seem necessary to call the members together for time-consuming discussion, the decisions were made by the chairman; the committee convened only to be informed as to what had been done...."

The four-year B.Chem. program for chemists specified a total of 142 credit hours, with fifteen courses in chemistry (six of them in analytical chemistry), plus mathematics, physics, miner-

> *. . . the engineers were also confident that [a five-year program] would fail, and that even if it didn't, the College of Engineering would have to deal with only a few new students....*

Above: "Mech Lab" in the early 1930's.

Below: Freshman chemistry lab at Baker in the 1930's.

alogy, engineering drawing, economics, and two courses in chemical engineering: industrial chemistry and Chemistry 710, which included the infamous Chemical Engineering Laboratory with its emphasis on report-writing. All for a degree in chemistry!

The aspiring chemical engineers endured the same burdens as the chemists during their first two years, then assumed even heavier ones later on. Engineering courses in mechanics and materials of construction, heat and power engineering, mechanical laboratory, industrial organization, electrical engineering and machine design were added on top of the chemistry curriculum. The chemical engineering students also had the privilege of taking the chemical engineering laboratory course and a course in chemical plant design simultaneously in the spring of their fifth year. The total credit requirement was 179, of which a grand total of eight were elective.

This was essentially the curriculum your writer unwittingly signed up for on entering Cornell in the fall of 1937.

To the surprise of everyone but Dr. Rhodes, the new program flourished. Understandably, only a few of the B.Chem. holders completed the five-year program: many didn't want to, and a number of those who did were "terminated" by Dr. Rhodes after their fourth year. But some did see it through: three in 1933, five in 1934, two in 1935, and seven each in 1936 and 1937. The numbers continued to grow after the establishment of the School of Chemical Engineering (see Appendix E).

The first curriculum in chemical engineering was admittedly unsatisfactory, for in the 179-hour program there were only three courses in the major field. This was all that could be included in view of the requirements of the College of Engineering and the College of Arts and Sciences. Some of the courses in chemistry and mechanical engineering were considered by Dr. Rhodes to be "less than useful", but they could not be deleted because professors of both chemistry and engineering persisted in their belief that chemical engineering was a hybrid and not a distinct discipline. Moreover, as Dr. Rhodes wrote later, the three chemical engineering courses were about all he could handle by himself.

In 1935 Dr. Charles C. Winding was hired as an instructor. The course Unit Operations was divided into a lecture and a laboratory course and both were moved from the fifth to the fourth year. A few required courses such as Optical Chemistry (Spectroscopy) were deleted, reducing the credit requirement to 170. Major changes in the curriculum, however, had to wait until after the establishment of the School of Chemical Engineering in 1938.

*To the surprise of everyone but Dr. Rhodes, the new program flourished.*

# 2. The Early Days of the School

The establishment of the School of Chemical Engineering in 1938 did not significantly affect the students already in the program. In fact, they were largely if not completely unaware that anything had happened. Classes continued unchanged in Baker, Sibley, and the other old familiar buildings. A new faculty member did appear, Assistant Professor Oscar J. "Och" Swenson, who taught chemical process design to the fifth-year students, but otherwise life went on as happily and as miserably as before.

A fundamental change had occurred, however, in that Chemical Engineering was now entirely in the College of Engineering. Professors Rhodes and Winding were no longer members of the faculty of Chemistry, and their continued presence in Baker Laboratory was not uniformly appreciated by their former colleagues. Students entering in 1938 and thereafter were also made aware of the new situation: their applications were to the School of Chemical Engineering, not Chemistry, and the five-year program led to only one degree, Bachelor of Chemical Engineering, not two degrees as before. The sole member of the School's admission committee was Director Rhodes, who typically would read some two hundred applications, admit perhaps one hundred students, and expect one-third of them to complete the course. "Shake hands with the person on your left and on your right," he would tell the new freshmen at their first meeting. "Only one of you three will graduate." Quite a welcome for a nervous, insecure matriculant!

World War II began in Europe in September 1939 and the United States became increasingly, though unofficially, involved in assisting Great Britain and its allies. The effect of this on the lives of Cornell chemical engineering students was minimal. Behind the scenes, however, plans unrelated to the war were being made, and their implementation would have a very large effect. Director Rhodes and Dean Solomon Cady Hollister began negotiating with Franklin Walter Olin, a member of the Cornell class of 1886 and president of Olin Industries, to raise funds for a new building to house the growing School of Chemical Engineering. They were persuasive, and in 1940 Mr. Olin gave $685,000 in memory of his son, Franklin W. Olin Jr., a Cornell civil engineering graduate of 1912. A grant of $29,000 from the University covered the overrun in the cost of the building; another grant of $250,000 from the Parmalee Fund was used to purchase laboratory equipment, lights, office furniture, and other necessary items. The unit cost of the 100,000-square-foot building was therefore $9.64 per square foot.

This was the first new engineering building at Cornell in

*Franklin W. Olin*

over thirty years. Although in the 1920's and 1930's several different plans for a new engineering complex at the north end of the campus had been developed, the economic depression of the 1930's put all such dreams on hold. Now the trustees decided that the engineering complex would be built at the south end of the campus, with Olin Hall as the first unit. The new building was located on what was known as Sage Green, where women students had once practiced archery and exercises in aesthetic dancing. The architects— Shreve, Lamb and Harmon of New York City, the designers of the Empire State Building— worked closely with Director Rhodes in designing the interior of Olin Hall to his specifications. Professor Swenson also made useful contributions to the project.

Director Rhodes had boundless energy and found time, even then, to run the Unit Operations Laboratory. In 1940–41 the equipment, in the basement of Baker Lab, was far from impressive; no money had been spent on it in years and much of it functioned poorly, if at all. What it was supposed to do was often a mystery as well— I recall watching the still or the evaporator, wondering what it was we were supposed to see and, more important, to write about. Even so, I was better off than some: several of us were over six feet tall and we would crowd together in front of the shorter students— in a friendly way, of course— to keep them, as far as possible, from seeing anything. Dr. Rhodes designed the experiments, helped get them running, assembled the week's data, calculated all the derived quantities, and had the results mimeographed and issued to us as the basis for our reports. He made graphs and equipment drawings, wrote descriptive notes, and edited all the reports with lengthy critical comments. Often he wrote more as criticism than the student had written in his entire report.

He also used his notorious grading system: one grade from zero to 100 for technical content, and one for style and presentation. The two grades were multiplied together and divided by 100 to give the final mark— thus, 70 and 70 gave a 49, and zero and 95 gave a zero. Late reports weren't accepted at all. Nearly always, the report had to be rewritten and resubmitted; an unsatisfactory revision meant an even lower grade, and a second unsatisfactory attempt meant a zero for the report. Even a zero-grade report had to be revised until acceptable, or one failed the course. Over two terms, there were ten multipart experiments, each with its own report; for this the student was rewarded with one hour of credit each term. It was a true if painful learning experience; it not only changed one's style of writing, but also the way one thought about things, and it showed us lazy students how much could be accomplished when standards were set very high.

Construction of Olin Hall began in April 1941. Founda-

*Bill Robinson '41 (at top) and F. S. "Pete" Hathaway '41 (below) were studying for their mineralogy final when these photos were taken in January 1940.*

*The two grades were multiplied together and divided by 100 to give the final mark— thus, 70 and 70 gave a 49, and zero and 95 gave a zero..*

*Above: The site of Olin Hall— Central Avenue in the 1920's.*

*Below: Excavation, and the discovery of bedrock.*

*Left: Olin Hall under construction in 1941.*

*Left: The dedication ceremony at the entrance to Olin Hall. Director Rhodes is at the podium and Dean Hollister is seated to his left.*

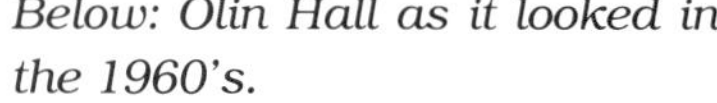

*Below: Olin Hall as it looked in the 1960's.*

tions were dug down to bedrock, the discovery of which necessitated shortening the north end of the front wing of the building by about seven feet. Structural steel was put up during the summer, with the last load arriving just in time to escape wartime restrictions on the use of steel for domestic construction. The building was largely completed in May 1942, and after the Commencement exercises that month, the last eighteen of the Chemical Engineer degrees were handed out in the sparsely furnished lounge. In September the School moved from Baker to Olin Hall; some classes were still given in Baker during the fall of 1942, but by February 1943 the transfer was complete.

It's hard to say who were happier— the chemists or the chemical engineers.

The interior design of the Olin Hall, with cinder-block walls, concrete floors, and windowless lecture rooms, was functional and effective, if not aesthetically impressive. It aroused few objections. The exterior appearance, however, generated a loud outcry from Cornell alumni who complained that the new building clashed with Willard Straight and Myron Taylor Halls; that it was too modernistic; that it departed from "traditional Cornell architecture," and, worst of all, that it looked like a building for chemical engineering— which, of course, is exactly what it was.

In 1941 Director Rhodes was named the Herbert Fisk Johnson Professor of Industrial Chemistry. In 1942 Professor Clyde W. Mason was lured from the chemistry department with promises of spacious new laboratories for his specialty, chemical microscopy. He became a member of the chemical engineering faculty and for many years taught chemical microscopy and materials of construction.

The materials course was a new one in 1941, replacing an unsuitable course offered by the mechanical engineering school. (See the 1937–38 curriculum in Appendix C.) Other curriculum changes in 1941 included increasing the mathematics requirement from two five-hour courses to four three-hour ones, with the addition of differential equations; moving Chemical Engineering Technology (then the first course in chemical engineering) to the third year; adding Chemical Engineering Computations to the fifth year; dropping the mechanical engineering course called Industrial Organization; and adding the course Plant Inspection. This still left a large number of courses in analytical chemistry, drawing, German, and mineralogy as required subjects in the curriculum. The total credit requirement for the B.Ch.E. degree was 182, up considerably from the requirement in 1937–38 .

Then on December 7, 1941, bombs fell on Pearl Harbor and everything changed.

*. . . by February 1943 the transfer was complete. It's hard to say who were happier— the chemists or the chemical engineers.*

# 3. Wartime and the Postwar Period

Once the United States was officially at war, increasing numbers of students withdrew to enter the armed forces. At first, engineering students in good standing were deferred under the Selective Service Act, but as time went on, deferments were less and less frequent, and by 1943 there were few able-bodied men students left in Chemical Engineering. At this time the Navy instituted what was called the V-7 (later V-12) program, in which engineering students who had finished two years of college could enter the Navy as midshipmen and continue their studies to complete a total of eight terms of academic work to qualify for the newly designated degree of Bachelor of Science in Chemical Engineering. In operating this program, Cornell accepted its own students and also students from Princeton, the Newark College of Engineering (now the New Jersey Institute of Technology), and elsewhere.

To accommodate the eight-term requirement and also permit the midshipmen to attend their Navy classes and mandatory drill, all electives and some less essential professional courses were eliminated from the chemical engineering curriculum. Everything was condensed and accelerated. Some chemical engineering courses were offered in the summer of 1942; in 1943, 1944, and 1945 all chemical engineering courses were offered every term, summers included. Holidays were Christmas Day and the Fourth of July; otherwise, instruction proceeded continuously. End-of-term breaks were two days long.

The effect on enrollments is shown in Appendix E. Classes were large the first two years— thirty-eight degrees were granted in 1943 and eighty-four in 1944, but in 1945 the number was twenty-two and in 1946 it was only six. Of the sixty-five V-7 or V-12 students who received the four-year B.S.Ch.E. degree, forty returned after the war for the fifth year and their B.Ch.E. degree.

During this period, Olin Hall provided the Navy with offices and classrooms for such subjects as map reading and navigation. The Navy also set up a one-year postgraduate program to train naval officers for liaison work with manufacturers of explosives. The curriculum, which led to the degree of M.S. in chemical engineering, included courses in both interior and exterior ballistics. These were both taught by Dr. Rhodes, although he had no experience in either subject and, for interior ballistics, no good textbook was available. Instituted in 1943, the program continued until 1947. From 1943 to 1946, thirty-five officers entered the program, twenty-seven of whom received the M.S. degree. In the fall of 1947 the final twelve of these officers had to suffer through the Unit Operations Laboratory course, with lengthy reports

*Holidays were Christmas Day and the Fourth of July; otherwise, instruction proceeded continuously. End-of-term breaks were two days long.*

*Right: Central Avenue in the 1940's. (Olin Hall is at the corner of Central Avenue and Campus Road.)*

*Below: A 1944 photo showing, left to right, Theodore S. Williams (M.S. '44), Lucy Broadhead, and Marion Besemer outside Olin Hall. Lucy is remembered as the staff member to whom students turned for help with their curricular problems.*

graded by Professors Rhodes and Smith; they enjoyed it even less than did the regular chemical engineering students.

Wartime saw the first women chemical engineering graduates: Beatrice Noback (Robbins), B.S.Ch.E. '45 and Inez Leeds (Moselle), B.Ch.E. '46. A few graduate students, some with the rank of instructor, helped with the laboratory courses and, at times, with the lectures.

Despite their enormous teaching loads, the faculty members found time to do other things. Professor Rhodes was a consultant to the Materials Division of the War Production Board and a member of the committee for the Office of Production Research and Development. With Professor Swenson, he taught some Engineering Science Management War Training Program (ESWMT) courses in Corning and Buffalo. Professor Winding was a consultant to the Rubber Reserve Corporation on synthetic rubber manufacture.

The war ended in September 1945, too late to affect enrollments for 1945–46. Professor Swenson resigned from the faculty to go into farming; later he became a consultant on sugar manufacture. Sensing that enrollments would soon rise dramatically, Professors Rhodes and Winding began looking for new faculty members, and quickly hired three assistant professors:

Julian C. Smith and Robert L. Von Berg from Du Pont, and Herbert F. Wiegandt from the Armour Research Foundation. In the area of metallurgy, Peter E. Kyle was hired as a professor, and Malcolm S. Burton as an assistant professor. (For a discussion of metallurgical engineering in the School of Chemical Engineering, see Chapter 5.) Burton, Kyle, Smith, and Von Berg began teaching in October 1946; Wiegandt came in January 1947. While Director Rhodes was away for the summer of 1946, Professor Winding also hired Frank Maslan for a one-year period. On his return, Rhodes was greatly unimpressed by Maslan's qualifications and gave him the title of acting assistant professor. Frank was drafted after spending only a few days in Ithaca, but returned in February for the spring term of 1947. He was reappointed for the academic year 1947–48; was asked to resign in January 1948; refused to do so; and, despite Rhodes' efforts and objections, served out his full one-year contract. He was not reappointed.

Frank Maslan had more than his share of mistreatment during his brief stay at Cornell. A graduate student known as Big John became upset with him for some reason and resolved to "get even." Soon after that, Frank received a mysterious letter from Alabama informing him that the forty mules he had ordered would be shipped without delay. Frank thought nothing of it until other similar letters appeared. Finally, a telegram arrived saying something like "MULES FORWARDED; WILL ARRIVE ITHACA MARCH 20." Sure enough, on March 20 he got a phone call. "This is the station master. Come down here right away and get your [expletive] mules. They're driving us crazy!" Professor Thorpe, who was then a graduate student sharing an office with Big John, relates that he asked Frank what the phone call was about and received a noncommittal answer. Frank very reluctantly went to the station to find— of course— no station master and no mules. Meanwhile, Big John was rolling on the floor of his office, doubled up with laughter. Ray Thorpe apparently enjoyed the joke also; at least, it made an indelible impression on his memory.

The veterans did return, in even greater numbers than expected. The enrollment in 1947 was 442, more than double that of 1946, and of these, 305 were veterans. Teaching loads increased proportionately, straining facilities and faculty to the limit. Many things were in short supply: sugar, appliances, cars, and, most of all, housing. Students and new faculty members had to live in all manner of places, mostly unsatisfactory; some students were even housed on upper floors of Olin Hall. Their opinion of these quarters was epitomized by the phrase scratched on the windowsill of Room 246, which proclaimed for years afterward, "WE HATE THIS PLACE!"

The 1947–48 curriculum (Appendix C) differed considerably from prewar curricula. German, mineralogy, and advanced

*The Unit Operations Laboratory, as pictured in the program for the dedication of Olin Hall. The double-effect evaporator is in the foreground.*

quantitative analysis disappeared. Chemical Engineering Thermodynamics replaced the chemistry thermodynamics course; credit for the Unit Operations Laboratory course was increased to three hours per term. Newly required courses included Chemical Engineering Stoichiometry and psychology or public speaking in the second year, and Science in Western Civilization (History 165, 166) in the third year. This last was an excellent course taught by Professor Henry Guerlac. Credit hours for the entire curriculum totalled 190, of which twenty were elective.

March 1946 saw the first issue of *Olin News*, a mimeographed publication written sporadically by Dr. Rhodes to communicate with the students. A similar publication, the *Cornell Chemical Engineer*, was sent on occasion to the chemical engineering alumni. There were only a few issues of the *Cornell Chemical Engineer*, however, and by July 1952 the *Olin News* was directed entirely to the alumni and not to students. (It became the *Olin Hall News* in May 1961, to avoid confusion with publications from the new John M. Olin Library.) Early issues dealt with such topics as elections to the Chemical Engineering Student Council (an advisory group which died in 1952 from lack of student interest); the decoration of the Olin Hall Lounge, funded by the Brunson family in memory of their son Robert, B.Ch.E .'43, who was killed in World War II; student criticism of teaching methods, as requested by Director Rhodes; parties in Olin Hall, with many songs and (even) beer. A student honor society, Pros-Ops, was started in October 1947. From March 1946 until February 1948, all examinations were given on the honor system, but increasing complaints by non-cheating students put an end to this experiment and examinations were once again proctored by faculty members or graduate assistants.

In the fall of 1947 the school officially became the School of Chemical and Metallurgical Engineering. In 1948 James L. Gregg came from Bethlehem Steel to be a professor of metallurgy, and in 1949 Dr. Jay E. Hedrick of the Shell Development Company was named a professor of chemical engineering. Professor Hedrick's speciality was chemical engineering economics. In 1949 Burton, Smith, Von Berg, and Wiegandt were all promoted to the rank of associate professor on the strength of a simple recommendation by Dusty Rhodes. This gave them tenure, which had been discussed by the faculty and trustees for the preceding ten years, but which had been made official only in January 1948. At least one of us didn't know he had tenure until several years later.

The number of applications from returning veterans decreased and those from nonveterans increased. Cornell, though greatly changed from the prewar years, began to return to normal.

# 4. The 1950's and the End of the Rhodes Era

By 1951 the postwar student bulge was almost over. Freshman admissions remained high, from one hundred twenty to one hundred fifty each year, but the number of B.Ch.E. degrees awarded, which had reached a peak of seventy-four in 1950, dropped precipitately to twenty-five in 1953, in part because Selective Service took many students for duty in Korea. The number of baccalaureate degrees rose again, fell to a low of twenty-two in 1955, and then, with some fluctuations, averaged close to thirty-nine for the next twenty years (see Appendix E). The demand for chemical engineering graduates in the burgeoning chemical industries was enormous; typically, one hundred fifty to two hundred companies, each wanting one to ten chemical engineers, would interview the fifteen to twenty-five seniors available that year.

Except for the Korean war, it was a peaceful time of growth and consolidation. The curriculum continued to evolve: Chemical Engineering Stoichiometry became Introduction to Chemical Engineering, psychology was dropped, and Applications of Statistics became a required course; but there was no fundamental change. The total credit requirement for the five years was lowered from 190 to 179. New faculty members were added: Raymond G. Thorpe (M.Ch.E. Cornell '50) came in 1951; Peter Harriott (B.Ch.E. Cornell '49, Sc.D. M.I.T. '52) came from General Electric in 1953; and Robert K. Finn (Ch.E. Cornell '42, Ph.D. Minnesota '49) came from the University of Illinois in 1955 to develop courses in biochemical engineering. In 1953 Professor Hedrick became assistant dean of the College of Engineering, with a half-time position in Chemical Engineering; he returned to full-time teaching in 1956. In 1956 Dr. Robert York of Monsanto was appointed the Socony-Mobil Professor, an industrially supported position with a then-generous salary of $12,000 per year. (In 1950–51 Director Rhodes' salary was $10,000.) There were several faculty additions and changes in the Division of Metallurgical Engineering as well (see Chapter 5). In 1958 Clyde Mason was named the Emile M. Chamot Professor of Chemical Microscopy.

One after another, new engineering buildings appeared across the road from Olin Hall as the College of Engineering moved from the north to the south end of the campus.

Research had been part of the chemical engineering program from the earliest days— chemical engineering had become a field in the Graduate School in 1931, and the first two Ph.D.'s were awarded in 1932— but research had always been very limited. Now it began to increase. Professor Harriott started a research program in catalysis, and Professor Finn began one in biochemical engi-

*The demand for chemical engineering graduates in the burgeoning chemical industries was enormous.*

16

*A faculty meeting in 1955. Left to right: Charles C. Winding, Fred H. Rhodes, Clyde W. Mason, Julian C. Smith, and James L. Gregg.*

*The acceptance, in 1955, of a $60,000 gift from the Socony Mobil Oil Company to establish a professorship in chemical engineering. Present are Director Rhodes (at left), William M. Holaday, director of Socony Mobil Research Laboratories (center), and Assistant Dean J. Eldred Hedrick.*

*Right below: "Synthetic gravel" and its developers, Professors Benjamin K. Hough (at left) and Julian C. Smith.*

*Below: Professor Clyde Mason at at senior class banquet, about 1953.*

neering. Professor Wiegandt invented a process for the desalination of sea water by freezing out ice through direct contact with a refrigerant. In the early 1950's Professor Smith received much publicity for his invention, with Professor B. K. Hough of Civil Engineering and under Army sponsorship, of a chemical method of stabilizing wet soils, a method that might also be used in the manufacture of "synthetic gravel." (Neither technique proved useful in the field.)

In 1956 Director Rhodes was granted a year's leave of absence (his first ever), and in 1957 he retired. An era ended for the School. A party was held in Ithaca for him and Mrs. Rhodes on June 8, 1957, and a testimonial dinner attended by many visiting dignitaries was given in the Plaza Hotel, New York City, on October 28. At this affair plans were announced to establish, largely through alumni contributions, the Fred H. Rhodes Professorship. After this, Professor Emeritus Rhodes spent much of his time away from Ithaca; he interfered not at all with the administration of the School. He served as an alumni trustee of the University from 1958 to 1963. In 1964 he moved permanently to De Land, Florida, and was active in civic affairs there until his death in 1976.

Dusty Rhodes was, as they say, quite a guy. As a young man he not only had a dream of a School of Chemical Engineering; he had the toughness, energy, and political skill to make it a reality, almost single-handedly. He had extraordinary powers of concentration: he did what he was doing, to the virtual exclusion of everything else. While in Ithaca, he worked seventy to eighty hours a week for the School; in the summer he would disappear, incommunicado, to a camp in Quebec province twelve miles beyond the end of the road, accessible only by portage or bush plane. He loved to master new subjects, then give a course or write a book about them. He published *Technical Report Writing* in 1941 and *Elements of Patent Law* in 1949; in the mid 1950's he wrote a text, *Applied Statistics in Chemical Engineering*, which was never published. I recall that for the three-week period while this was being written he did nothing else— his duties as director were carefully and completely ignored. (A list of books published by members of the faculty of the School of Chemical Engineering is given in Appendix G.)

Around 1953 he tried to create an "instant tradition," decreeing that henceforth all chemical engineering students would wear an old school tie of his own design, a garish affair with stripes of red and black and gold. For a week or two, the students went along with the idea, but soon there were defectors and by term's end the "tradition" had died. The graduating seniors did give him a vest made of the same material, which he proudly wore on state occasions. (See the frontispiece.)

His likes and dislikes were intense. For one thing, he hated

*Below: Everett Kelm of the Corning Glass Company and Professor Julian Smith with the all-glass column in the Unit Operations Lab in 1950.*

*Right: Dusty Rhodes giving his last lecture as a member of the School faculty.*

*Below: The Fred Hoffman Rhodes Recognition Dinner held October 28, 1957 at Hotel Plaza in New York City.*

meetings, especially discussion meetings. As a faculty, we met only twice a year to take academic actions, which usually meant approving what Dusty had already decided to do. He was arbitrary, even hasty, in many of his decisions. He had little patience with incompetence and none at all with laziness. On occasion he deliberately provoked uneasiness and fear, sometimes bordering on terror, among the students. A few cordially disliked him, but to the majority he was a generally sympathetic father figure who listened to their problems and complaints and helped them through difficult times. More than a few idolized him.

Dusty loaned or gave a lot of money from his personal funds to needy students and young faculty members, and organized a student loan fund through solicited donations from alumni. He worked unceasingly to get to know his students individually, bring out their very best abilities, get them good jobs and good salaries, and keep in close touch with them after they graduated. He did pinch pennies, though, "until they squealed," as Dr. Winding said later. Each year he prided himself on turning back some part of the University's meager budget allocation for the operation of the School, which meant that administrative expenditures and faculty niceties were minimal. There were few frills. In those days research was not well supported and there were hardly any externally sponsored projects; what little was done had to be carried out as cheaply as possible. Dusty was not opposed to research— he even encouraged it, mildly— but teaching, especially undergraduate teaching, was by far the most important thing to him. He once wrote, "The prime responsibility of the University is the training of undergraduates— everything else should be of secondary consideration and not permitted to interfere with undergraduate instruction." All in all, he was a remarkable figure: a great builder, an effective if sometimes aggravating administrator, and a truly inspiring teacher.

*A few cordially disliked him, but to the majority he was a generally sympathetic father figure. . . .*

# 5. The Rise and Fall of Metallurgical Engineering

For seventeen years, a five-year degree program in metallurgical engineering was offered by the School of Chemical Engineering. The Division of Metallurgical Engineering was authorized in 1946, and in 1947 the school was renamed the School of Chemical and Metallurgical Engineering. The first faculty members, Peter E. Kyle and Malcolm S. Burton, were hired in 1946 as professor and assistant professor, respectively, of metallurgy. In 1947 the Francis Norwood Bard Professorship of Metallurgical Engineering was established, and Professor Kyle was appointed to that chair.

The new program had been created by (who else?) Fred H. "Dusty" Rhodes, who had had such an undertaking in mind for a long time. For three years, around 1936, he had taught a course in furnace metallurgy, but gave it up to concentrate on organizing the School of Chemical Engineering. Instruction in metallurgy and related subjects, however, was nothing new at Cornell. Assaying of ores had been taught in the Department of Chemistry from 1868 to 1916, and at various times the requirements for a bachelor's degree in chemistry included courses in assaying, metallurgy, and mineralogy. Metallography was offered, also in Chemistry, by Professors Chamot and Mason. The Schools of Mechanical Engineering and Civil Engineering sporadically offered courses in metallography and metal-working, and foundry work was required of all mechanical engineers from 1880 on. There was even a metallurgical engineering option in mechanical engineering in 1939–40. It was only after World War II, however, that the demand for metallurgical engineers was considered great enough to justify a separate degree program.

Professor Mason, in consultation with others on the faculty, developed the first metallurgical engineering curriculum. It stressed physical metallurgy (fabrication, structure, and physical properties) rather than process metallurgy (ore beneficiation, smelting, and refining), and was designed to prepare students for work in the metal-working and fabrication industries. The ties with chemical engineering were clear: the first-year courses were almost identical with those for freshman chemical engineers, and many of the later courses were shared by both fields. A novel concept—unit processes—was introduced into the metallurgical engineering curriculum. The curriculum for 1957–58 is shown in Appendix D; it included seven required courses in metallurgy. The total credit-hour requirement was 171.

The first metallurgical engineering students, twelve freshmen and fifteen sophomores, enrolled in 1947–48. The old foundry, which had belonged to Mechanical Engineering, was made

*The new program had been created by (who else?) Fred H. "Dusty" Rhodes. . .*

part of the School, and was extensively rehabilitated by Professor Kyle with a grant from the American Foundrymen's Association. Chemical and Metallurgical Engineering also acquired the foundry support staff, Morris Harper, Dennis Joyce, and Ralph Hodges, who were given the title of Instructor Technician. They supervised the laboratory sections in foundry practice and other courses and handled a few recitation sections in the earlier courses in metallurgy. In 1948 James L. Gregg, after many years at Bethlehem Steel, was brought to Cornell as Professor of Metallurgical Engineering.

Professor Kyle also obtained grants from Bausch and Lomb for a first-class metallography laboratory and from Air Reduction for a laboratory for metal cutting and welding. These were set up in Olin Hall.

The first nine B.Met.E. graduates received their degrees in 1951. Thereafter, the annual number of degrees fluctuated between two and six, with no sign of a long-term increase. Cornell was late on the scene with a program of this kind, pretty much limited to conventional metals, at a time when new metals and other new materials of construction were rapidly being developed. A few master's and doctor's degrees were awarded, but research— seen as important by the metallurgical engineering faculty if not by Dr. Rhodes— did not begin to flourish until the 1960's. Even so, except during the "bad year" of 1958, the metallurgical engineering graduates readily found well-paid employment, not only in traditional metal-processing and fabrication companies, but increasingly among chemical manufacturing firms.

One anecdote from this period, reported in Morris Bishop's *A History of Cornell*, seems too good to omit. In those years, "Engineers' Day" was a sort of open house during which students and faculty members set up displays around the engineering

*Scenes at the Cornell foundry, predecessor to a laboratory for metallurgical engineering. The foundry was extensively renovated in the 1950's.*

campus demonstrating various scientific and engineering phenomena and curiosa. On one such occasion, Professor Mason stood by a microscope, inviting the public to look at a slide of steel structure. Dusty Rhodes coached an eight-year-old girl, who stepped up, begging and whining for a look. When she was finally permitted to peer through the microscope, she chirped, "Oh, see the graphitic precipitation of carbon in the grain boundaries!"

Professor Kyle, although (according to Rhodes) a cooperative and loyal member of the faculty, did not live up to his promise as a contributor to the development of the metallurgical engineering curriculum. His chief interest was foundry practice. He maintained his membership in Lesselles Associates, a Boston consulting firm, and spent a great deal of his time on consulting work. In 1954, at a time when he was beset with domestic problems, his removal was requested by the other members of the metallurgical engineering faculty, in a petition— virtually an ultimatum— to Director Rhodes. Reluctantly, for he thought that with tact and understanding the matter could have been peacefully resolved, Dr. Rhodes asked for, and got, Professor Kyle's resignation. For a year the position was unfilled; then, in 1955, Dr. George V. Smith of the research division of U.S. Steel was hired as the Francis Norwood Bard Professor, and in 1956 became the assistant director of the School. In 1956 and 1960, respectively, Chester W. Spencer and Harry W. Weart were hired as assistant professors of metallurgy. In 1961 John B. Newkirk came as a professor of metallurgical engineering without tenure, as did Eraldus (Pete) Scala in 1962; also in 1962, Che-Yu Li, a Cornell Ph.D. in chemical engineering, was appointed assistant professor. From 1959 to 1962, Dr. Thor Rhodin was listed as an associate

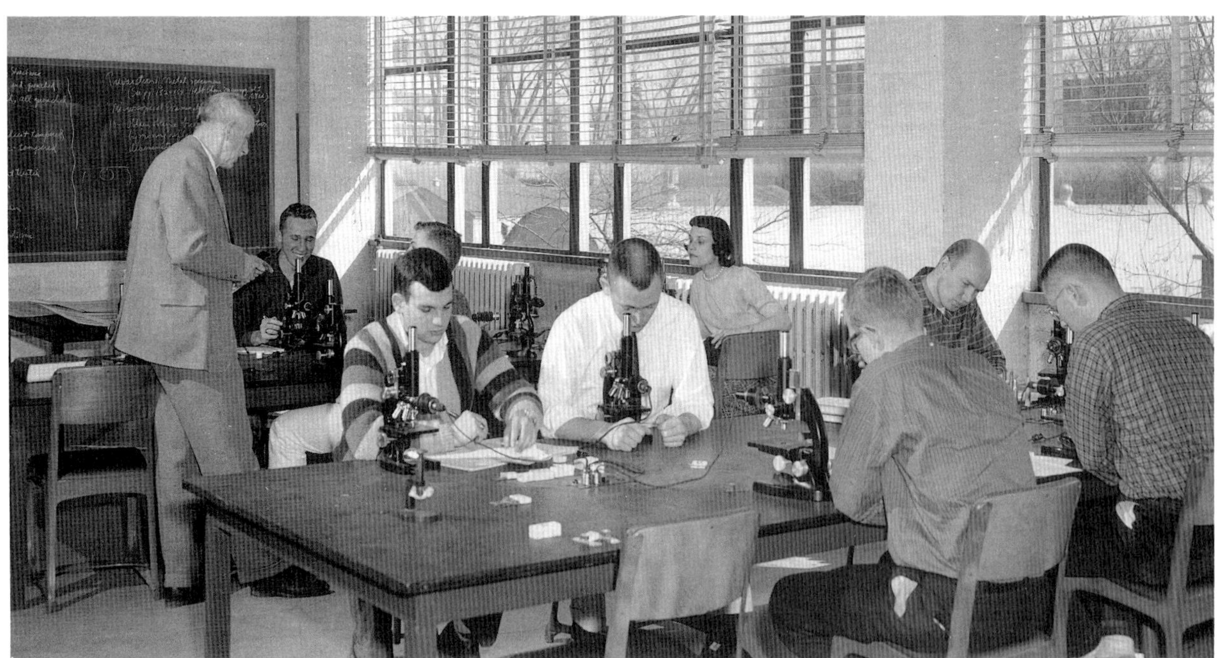

*Below: Professor Mason and his students in the chemical microscopy laboratory.*

professor of engineering physics and metallurgical engineering; he did not, however, have an office in Olin Hall.

In Olin Hall in 1962 there were nine faculty members in chemical engineering and eight in metallurgical engineering, plus Professor Mason, who served in both areas. The number of graduate students had increased to forty. In those days, since the fifth-year students had to carry out both a design project and an individual research project, each student was given office space in Olin Hall, a room usually shared with one or two others. Consequently, Olin Hall had become very crowded. When funds for Bard Hall of Metallurgical Engineering finally became available, the chemical engineering faculty grew increasingly impatient for their former colleagues to vacate Olin Hall. It took a long time— it wasn't until May 1963 that Bard Hall was finished and the metallurgical engineers moved out, giving back some 12,000 square feet of floor space. Professor Mason stayed with the chemical engineers. The school became once again the School of Chemical Engineering.

Dean Hollister retired in July 1959; his place was taken by Dale Corson, formerly chairman of the Department of Physics. This was the post-*Sputnik* era, in which there was an almost hysterical concern for improved training in science. Traditional engineering, especially empirical knowledge and industrial practice, was demoted by many to "trade school stuff." It was soon evident that under Corson, engineering would become more "scientific"; that is to say, more clearly based on physics. Dean Corson's announcement to the faculty of Metallurgical Engineering of his plans for their unit to become part of a new department, based on physics, caused much unhappiness. Professor George Smith resigned his position as head of the department, and for the next two years Professor Scala served in that capacity. In July

*Left: Presentation of the Francis Norwood Bard Professorship of Metallurgical Engineering. Here Bard is speaking at a dinner held in his honor on November 7, 1947, The others, left to right, are Peter E. Kyle, the first holder of the professorship; Dean Hollister; and Director Rhodes.*

1963 the Department of Engineering Physics and Materials Science was formed and put under the direction of Professor John Howe. This arrangement, merging faculty from three basically incompatible departments, lasted only a few years, after which separate departments of Engineering Physics and of Materials Science and Engineering were established.

The entire metallurgical engineering faculty left Olin Hall in 1963 and joined the new department, except for Professor Spencer, who resigned to go into industry. Thor Rhodin became Associate Professor of Engineering Physics; George Smith continued as the Bard Professor of Metallurgical Engineering. During the next few years, the curriculum changed a great deal as the emphasis shifted from metallurgy to materials science and from industrial practice to research. Soon the term "metallurgical engineering" disappeared, even from the sign on Bard Hall; after Mr. Bard's death, this was changed to read "Materials Science and Engineering". In 1965 Professor Newkirk was denied tenure and left Cornell to become department head at the University of Denver; soon after Professor Weart went to a similar position at the University of Missouri at Rolla. In 1970 Professor Gregg retired and Professors Scala and Smith resigned from Cornell. Of the Metallurgical Engineering faculty, only Professor Burton and Professor Li continued at Cornell.

Today little is left of the old curriculum. According to the 1987–88 Cornell catalog, *Courses of Study*, only one course with "metallurgy" in its title is currently offered by the Department of Materials Science and Engineering. Two courses in mineralogy are given by Geological Sciences, but courses in metallography are not offered anywhere in the University.

*Sic transit gloria curriculorum.*

> *Soon the term "metallurgical engineering" disappeared, even from the sign on Bard Hall. . . .*

# 6. The Winding Era— the Peaceful Years

In 1957 Charles C. (Chuck) Winding was appointed director of the School. He had been acting director the year before, when Dusty Rhodes was on leave, so there was no obvious change when he officially took over. Chuck had been strongly in favor of Rhodes' policies, and for the next four years things went along very much as they had before. Except during the brief recession in 1958, Cornell chemical engineering graduates continued to be much in demand by industry.

In 1958 Dr. Winding did make some changes in the basic program, in response to the pressure for a more "scientific" curriculum. Engineering drawing was reduced from four to three credit hours. Three courses— in accounting, electrical engineering, and chemical engineering equipment— were dropped. Reaction kinetics, advanced inorganic chemistry, and five hours of electives were added. The graduate program was also revised, so that advanced courses in mathematics, physics, chemistry, unit operations, and reaction kinetics were required of all graduate students.

The course Plant Inspection was terminated a few years later. This one-credit course had been established in the late 1940's to give fourth-year students a chance to see (and report on) nine or ten processing plants of various kinds— chemical, petroleum refining, coke production, even steel-making and sugar refining— all in a six-day period during spring vacation. Buses took the senior class and faculty chaperones to Rochester and Buffalo in some years, to northern New Jersey or the Philadelphia-Wilmington area in others. It was demanding on students and faculty alike. By the late 1950's the increased availability of summer jobs for undergraduates in chemical and petroleum plants had reduced the need for this kind of a course, and just about everyone heaved a sigh of relief when it was abandoned in 1961–62.

In 1956 Director Winding had started a six-year professional nonthesis master's program leading to the degree Master of Chemical Engineering, and in 1960 he instituted a "predoctoral honors program" for students interested in careers in teaching or research. More and more B.Ch.E. graduates were going to graduate school— some 30 percent of the class, compared with less than 20 percent in earlier years. They did well in graduate school, too; they were well prepared and generally more mature than the other graduate students. Often they received little or no academic credit for their fifth year of undergraduate study, however; the honors

*They [B.Ch.E. recipients] did well in graduate school, too; they were well prepared and generally more mature than the other graduate students.*

*Right: A 1947 meeting of alumni and faculty members during a Plant Inspection visit to the New York City area. Standing, left to right: George Baumann, Ray Van Swearingen, Bill Taylor, Jack Weikart, Bert Belden, Bill Schumacher, Sr., and Ed Ricker. Seated, left to right: Whitey Nelson, Harry Hilleary, Jr., Dusty Rhodes, Dave Hogin, Tom Buffalow, Robert Von Berg, and Chuck Winding.*

*Right: A 1957 classroom scene.*

*Below: The School's business office, 1957. Janet Ott is the secretary.*

*Right below: Two of a group of fifty high school teachers who attended a six-week Shell Merit Fellowship Seminar at Cornell in 1959. Professor Julian Smith is explaining a vacuum induction melting and casting unit.*

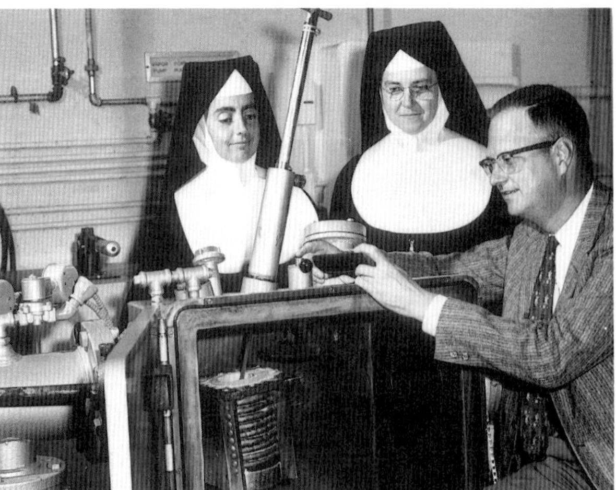

program was designed to remedy this situation as well as to provide even better preparation for graduate work. Students had to commit to the program during their sixth term; after approval, they were excused from courses such as Plant Design and the second term of Unit Operations Laboratory so that they could take advanced courses in mathematics and engineering theory. In the seventh term they had to demonstrate initiative in preparing for a two-year-long research project, and show they had the ability to cope with the graduate-level courses they would take during the fifth year. From 1961 through 1964, the annual enrollment in this program was about five students, most of whom did go on to doctoral studies at Cornell or elsewhere.

In 1958 Ferdinand Rodriguez received his Ph.D. degree from Cornell and was hired as assistant professor of chemical engineering. The next year Solomon Cady Hollister retired after twenty-two years as dean of the College of Engineering; his place was taken by Dale Corson, previously chairman of the Department of Physics. Soon there was a step change in the pace of research in the College, with a $4.1-million grant from the Ford Foundation and large grants for the Arecibo radar-radio telescope facility and the Materials Science Center. In two years the research budget of the College increased eight-fold.

Activity was evident in educational and curricular matters as well. In 1960 the College faculty established the Division of Basic Studies and proposed a common curriculum for all engineering freshmen and sophomores. It was to include one, or at most two, courses in chemistry, in effect destroying the chemical engineering program. Director Winding and his faculty spent tedious hours in meeting after meeting, arguing that chemical engineers are "different," that unlike other engineers, they must have a solid grounding in chemistry. Eventually, their views prevailed and chemical engineering students were allowed to elect a second chemistry course in the freshman year and to follow an "uncommon" program as sophomores. When all was said and done, the effect on the chemical engineering curriculum was fairly small, but Dr. Winding wrote that "it set us back at least two years in our efforts." After this change, applicants who wished to study chemical engineering were no longer admitted directly to the School, but only to the College, and were enrolled in the Division of Basic Studies for their first two years.

Dr. George F. Scheele, a recent Ph.D. from the University of Illinois, was named assistant professor of chemical engineering in the fall of 1961. In 1960–61 the courses in organic chemistry were moved to the third year; analytical chemistry was combined with physical chemistry and given in the second year, in recognition, at long last, that chemical analysis was of marginal importance to chemical engineers. The number of courses per term was

*Right: Students with a plant model (1969).*

*Below: Models prepared by students in Professor York's Plant Design course.*

*Right: Professor Rodriguez and a student measuring the elastic properties of collagen gels.*

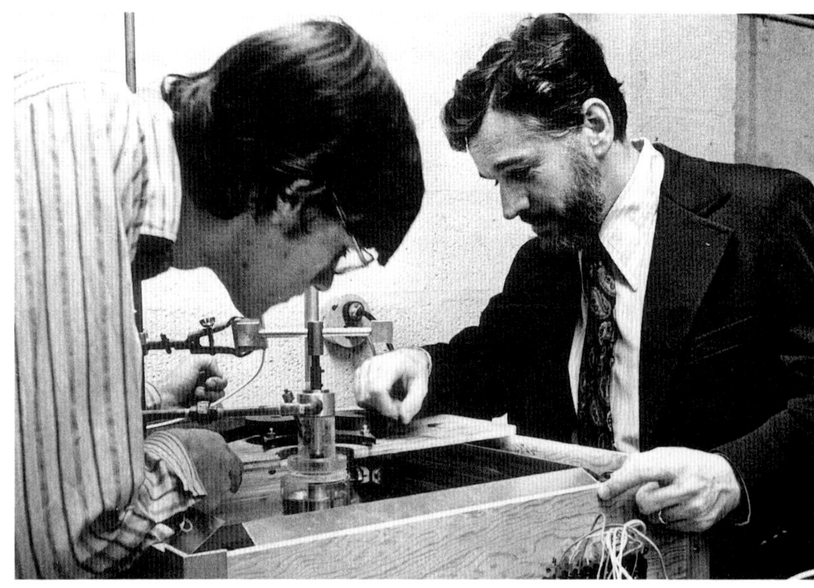

reduced to five, with a nontechnical course each term. The total credit requirement was 173, with thirty-six hours of electives and thirty hours of "liberal" courses. In 1962 a mathematics sequence with required training in the use of computers was introduced. Cornell had a computer in Rand Hall during the 1950's, but it was a slow, small-capacity machine used only by a few devotees. It wasn't until 1957 that a Chemical Engineering alumnus reported that he was involved professionally with computer usage. By 1962, however, it was clear that sooner or later, computers were going to change things in chemical engineering practice.

Plant Design, in those days, was taught in the fifth year as a two-semester course giving five credits each term. In the 1940's and 1950's the entire faculty had directed design projects, acting largely independently of one another; now the course was given entirely by Professor York. In 1962 he gave his students the choice of preparing conventional layout drawings or constructing a model of their proposed plant. They all chose models. Supplies were made available through grants from Procter and Gamble and M. W. Kellogg, and in 1963 a model shop was set up in the basement of Olin Hall. Typically, the students spent six weeks of the spring term building models.

About 1961, because of changes in the physical chemistry courses, it appeared for a time that Chemical Engineering might have to teach its own courses in that area. Laboratory benches were installed in Room 345 to make this possible if it became necessary— but it never did. Compromises were reached with the Department of Chemistry, and things went on much as before.

In 1963, after the metallurgical engineers moved to Bard Hall, a biochemical engineering laboratory was established for Professor Finn in the northeast corner of the first floor of Olin. In the basement, the Geer Laboratory of Plastics and Rubber—

*Robert York*

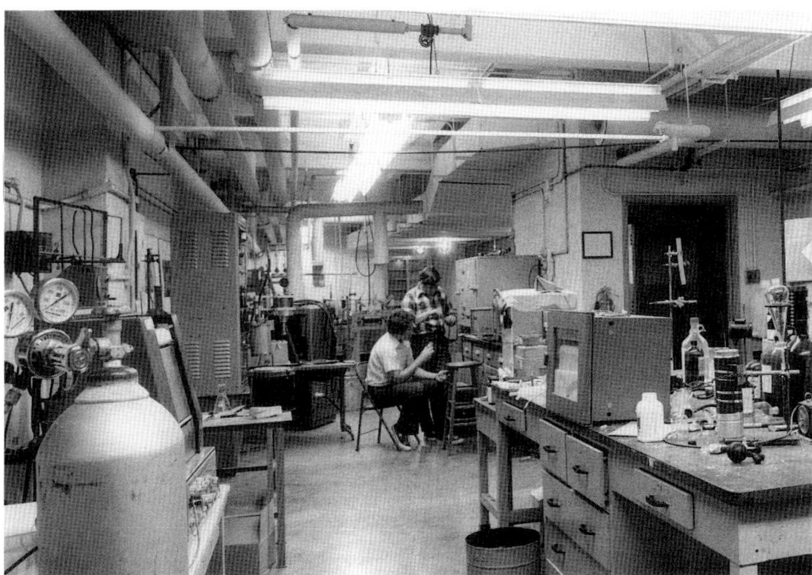

*Left: The Geer Laboratory of Plastics and Rubber.*

*Right: Chemical Engineering faculty members at commencement in 1964. Front row, left to right:: Professors Smith, Cocks, York, and Harriott. Back row: Professors Leinroth, Mason, Scheele, and Winding.*

established in 1951 through a gift of equipment from inventor William C. Geer of Ithaca—was expanded for a second time. The new Center for Applied Mathematics moved into some available space in Olin Hall.

Except for the problems with the "common" engineering curriculum, the seven-year period beginning in 1957 was a rather uneventful, peaceful time. In May 1963 Dr. Winding told the alumni, "No one seems to be mad at us and we're not mad at anyone else." Then, in the summer of that year, things began to happen. There were disturbing signs of future change. A committee under Professor Lloyd P. Smith of Physics studied the organization of the College; their "Smith I report" suggested that the traditional schools and departments were obsolete and should be replaced. Dean Corson became provost of the University, and Andrew Schultz, Jr. became acting dean of the engineering college. (Schultz was named dean in July 1964.) In November 1963 President John F. Kennedy was assassinated, and after that things were never quite the same. The peaceful days for Chemical Engineering were at an end.

*"No one seems to be mad at us and we're not mad at anyone else." Then, in the summer of that year, things began to happen.*

# 7. The Winding Era—
## the Turbulent Years

"The whole place seems to be ablaze, including the dispositions of most staff members," wrote Dr. Winding in the *Olin Hall News* of June 1964. The Policy Committee of the College of Engineering faculty had recommended a four-year Bachelor of Science program for all engineering undergraduates and a fifth-year "professional" program leading to the degree Master of Engineering. At the same time, a committee headed by Julian C. Smith of Chemical Engineering issued the "Smith II Report" on the reorganization of the College, recommending the establishment of twelve research departments based on scientific disciplines and ten "faculties" to do the teaching and handle the professional degree programs. Budget authority and faculty affiliations would be in the research departments; the traditional schools and departments would largely disappear. The college would be renamed the College of Engineering and Applied Science, in which Chemical Engineering would become Chemical Systems (a research department), and its professors would meet periodically as the "Faculty of Chemical Engineering" to discuss curricula or certify students for degrees.

Both proposals caused a lot of faculty unhappiness.

The Smith II Report was presented to the engineering faculty on May 14, 1964 and conveniently tabled. Except for some small changes made at the Sibley School of Mechanical Engineering, its recommendations were never implemented.

The proposal for a four-year B.S. program was another matter. Although the chemical engineering curriculum had been a five-year program from its inception, other engineering programs had been lengthened to five years only in 1946, and their faculties were not entirely convinced of the value of the longer program. Arguments for the four-year program stressed that Cornell's baccalaureate graduates in engineering who went on for advanced studies received no credit for their fifth year, that starting salaries for all bachelors-degree holders had become virtually the same, regardless of the length of their undergraduate program, and that the difference between the starting salaries for masters and Cornell's five-year bachelors, formerly small, had become substantial. Besides, an increasing number of students wanted to change fields after four years, to go into law, medicine, or business school. The proposal was debated at length, often acrimoniously ("educational hash" was one of the milder epithets applied to it), but it was approved by the College faculty on May 25 by a vote of 74 to 45. The five-year baccalaureate program was no more.

*Chuck Winding*

*Above: Two professors with students: Edwards (in the photo at left) and Finn.*

*Right: The Unit Operations Lab in the 1960s. Note the pan dryer (foreground) and the Votator (background).*

*Below: Professor Cocks in his microscopy laboratory.*

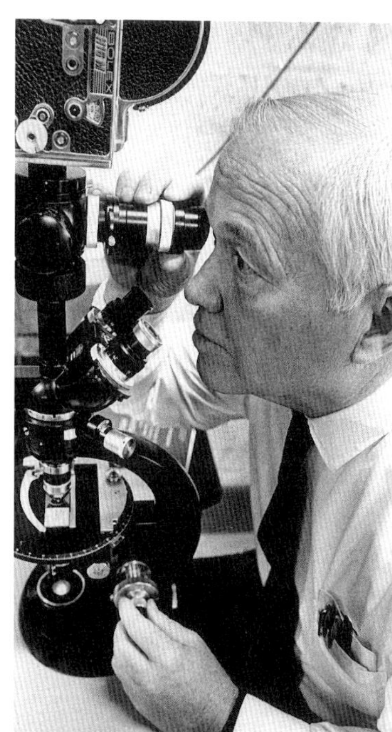

In Chemical Engineering, the four-year B.S. program was implemented with fewer difficulties than expected, and by 1965 Director Winding, in the *Olin Hall News*, was telling the alumni about its many advantages. Undergraduates had to have an academic average of 75 percent or more to be eligible to go on for the M.Eng. (Chemical) degree; even so, Dr. Winding expected about 90 percent of the seniors to continue. He also assured the alumni, "The B.S. degree does not mean the student is ready to accept professional responsibilities. We have no intention of recommending a four-year graduate for chemical engineering work." In 1965, 1966, and 1967 a large fraction (though not 90 percent) of the fourth-year students did continue for the M.Eng. (Chemical) degree. Some of the recent five-year B.Ch.E. alumni also returned for their M.Eng. degree, which they could get by attending Cornell for one term's work.

The Master of Engineering program was governed by the College's Graduate Professional Engineering Programs Committee, which was initially chaired by Professor Smith. In 1965 Smith left the School of Chemical Engineering to become director of continuing education for the College of Engineering; he did, however, continue to teach one chemical engineering course each term. In 1964 the School hired Dr. George G. Cocks from Battelle Memorial Institute as an associate professor of microscopy, and named Jean P. Leinroth (B.M.E . Cornell '41, Sc.D. MIT '63), who had many years of industrial experience, an associate professor of chemical engineering. Both were appointed without tenure. In 1966 Professor Mason retired and Professor Cocks took over his courses in microscopy and materials. Professor Leinroth helped develop the computer courses required of all freshman and sophomore engineers and gave a junior-level course in separations with emphasis on computers. Lemuel Wingard (B.Ch.E. '53, Ph.D. '65) was hired as an assistant professor for 1965–66, to work with Professor Finn in biochemical engineering. He later joined the faculty of the University of Denver. In 1966, Victor H. Edwards (Ph.D. Berkeley) was hired as an assistant professor in the biochemical engineering area. Also in 1966, Professor York had a heart attack and was on leave for part of the year.

During this period, the School had its first visiting professor, Kenneth Davis of Du Pont, who taught Process Control in the spring of 1966. Our first research associate was Dr. Hans Egle of Stuttgart, who worked with Professor Finn in 1964 on a USDA-sponsored project. Other research sponsorship during these years included NSF grants for an electron microscope and for the reaction kinetics laboratory, and Office of Saline Water support for Professor Wiegandt's desalination studies.

The 1967–68 curricula leading to the B.S. and M.Eng. (Chemical) degrees are shown in Appendix C. Taken together, they differ only moderately from the B.Ch.E. curriculum of 1957–58, and reflect the belief that most chemical engineering students would continue for the advanced degree. The Master of Engineering was to be the first professional degree; only there do we find courses in chemical process design. In the first and second years of the B.S. program, liberal electives have replaced English; organic chemistry has been moved to the third year and physical chemistry put in its place; quantitative analysis and public speaking have disappeared. In the third and fourth years, mechanics is a two-term course while materials and the unit operations laboratory have become one-term courses; a project laboratory has replaced the old fifth-year Senior Project; and a required course in reaction kinetics has been added. In the fifth year, now at the graduate level, the courses in process and plant design have been expanded from eight credits to twelve; the courses in electrical engineering and chemical engineering computations have been dropped, and courses in higher calculus and chemical process control added. The total credits required for the B.S .degree are 144; for the B.S. and M.Eng. (Chemical) degrees, 175. The requirement for the B.Ch.E. in 1957–58 was 176 credits.

Faculty turmoil over reorganization and the four-year baccalaureate program had largely ended by 1967, but the more serious and far-reaching student turmoil was well underway. The general atmosphere on campus had become increasingly uncertain after President Kennedy's assassination in late 1963. The Vietnam war was escalating, seemingly without limit. Student protests appeared, demanding— at first— such things as freedom from curfews and other campus rules. Hair grew longer, clothes sloppier. In 1966 a group of protesters gathered on the steps of Olin Hall to burn their draft cards. In 1967 and 1968 the protests became more ominous: there were demonstrations against ROTC, the war, the draft, racism, the industrial-military complex, society in general, and ultimately, authority and restrictions of any kind. Relatively few engineering students joined these activities, partly because they were deferred under Selective Service rules; in 1967 even graduate students in engineering were exempt. In 1968, however, deferments for graduate work ended, and enrollments in the Master of Engineering program dropped sharply. Some graduating seniors went into the Public Health Service to fulfill their military obligations, but for many students, sick of a war they couldn't in conscience support, a very real question was, "Should I go to Canada, or go to jail?"

Crusades for increased programs for minority students, especially Blacks, appeared. There was good reason for this. It's hard to remember how much racial discrimination there was in those days. Cornell's first Black chemical engineering graduate, Harry J. Green III, received his B.Ch.E. in 1963. His abilities, personal qualifications, and grades were all good and the industrial demand for chemical engineers was high, yet only one company— Corning Glass— was willing to consider him for a job.

By 1969, quite a few Black students were studying at Cornell; in overseeing their programs, the administration made decisions that appeared to many people both unclear and overly permissive. A breakdown in communications precipitated a series of confrontations. Finally, after a half-dozen Blacks were given a mild reprimand for a minor infraction, a group of Blacks responded to this and other perceived grievances by occupying Willard Straight Hall on Saturday, April 19. On Sunday, having been promised the reprimand would be lifted, they marched out again carrying guns and bandoliers, to the delight of photographers from *Newsweek*. Classes were suspended for a week. On Monday some eleven thousand people— faculty members and students— crowded into Barton Hall to hear unreassuring words from President Perkins; later that day over eleven hundred professors met and voted to sustain the reprimand. The Blacks threatened to destroy the University ("Cornell will *die*," said their leader over the radio). For twenty-four hours there was a real crisis. On Wednesday the faculty reversed itself and removed the reprimand, and things began to calm down. The endless meetings continued, however, leading finally to the organization of a Constituent Assembly to study restructuring the University. In Chemical Engineering a group of eighty students created the Chemical Engineering Forum, whose members conferred with the director, held meetings, and wrote a very good "open letter" to the alumni

*The mass meeting in Barton Hall during the campus disturbances of the late 1960's.*

asking for opinions on the organization, procedures, and course offerings of the School.

During the next academic year, 1969–70, the positions of University ombudsman and judicial administrator were established, as were the *Cornell Chronicle*, a new student code, and a trained police force. A University Senate consisting of faculty, staff, and students was created. In spite of all this, the campus atmosphere continued to be tense. There were sporadic protests, especially after the Africana Center was burned in the spring. For several weeks in April and May, the faculty and staff stood fire-watch at night in Olin Hall. A student "strike" in May was unsettling and disruptive, but ultimately ineffective. By then the Chemical Engineering Forum had been disbanded. The alumni reponses to the open letter had been tabulated, but "nobody knew where the results were." Student attendance at faculty meetings had dropped to zero, as students quickly learned how dull most faculty meetings are. After the strike, all Cornell students were given the option of receiving a grade of S (for Satisfactory) in subjects outside their major, without taking final examinations; about 15 percent of the chemical engineering students exercised this option.

In February 1969, David M. Watt, Jr., a Ph.D. from Berkeley, joined the faculty as an assistant professor. Dr. Philip G. Ashmore of the University of Manchester was a visiting professor in the fall of 1968, as was Dr. Philip Calderbank of the University of Edinburgh in the spring of 1969. In 1970, after twenty-seven years in Olin Hall, the Navy moved out. Quickly replacing it were new "tenants"— the Division of Unclassified Students, the Judicial Administrator's office, the Reading and Study Center, and the Guidance and Testing Center. In addition, the Center for Applied Mathematics and the University's telephone office continued to occupy space in an increasingly crowded building.

The pace of research had quickened a great deal in the years just past, but Director Winding's chief concern and emphasis was on undergraduate teaching. It was an unfashionable position at the time. Research and graduate studies took priority in the minds of the administration and much of the faculty, and the other engineering schools and departments were trying to outdo one another in building up graduate programs and attracting government-sponsored projects. Chemical Engineering was considered to be lagging far behind. In 1967 Dean Schultz appointed a visiting committee consisting of R. Byron Bird of the University of Wisconsin, William E. Hanford of Olin Industries, Richard R. Hughes of Shell, James Wei of Mobil, and Arthur Bueche of General Electric (chairman) to study the School and its future. They found the School to have an excellent undergraduate program, but to be out of touch with current research in industry and other universities.

They recommended increased emphasis on research; more industry support for the graduate program; more government-sponsored projects; and more applied mathematics and theory to be introduced in new or existing courses.

It was recognized that increased research would require additions to the faculty, and soon a search committee began looking for a new "distinguished professor." The committee members were Professors Rodriguez and Finn of Chemical Engineering, Professor Robert Hughes of Chemistry, and the dean of the Graduate School, Donald Cooke. A number of prominent chemical engineers were approached and interviewed. On Saturday, March 21, 1970, the dean called a meeting of the chemical engineering faculty to inform them that he had not only hired Dr. Kenneth B. Bischoff of the University of Maryland to be the Walter R. Read Professor of Engineering, but had also named him director of the School of Chemical Engineering to replace Chuck Winding. The faculty protested that they had had no say in the decision, but the dean was adamant. The change was made. Dr. Winding, who was on vacation in the Bahamas at the time, was informed of the developments by telephone. In August 1970 Dr. Bischoff assumed his duties as director of the School.

Dr. Winding accepted the new situation with admirable equanimity. From his communication to the alumni in the *Olin Hall News*, one could learn nothing of the bitterness generated in the chemical engineering faculty by the change of directors. He continued his teaching, devoting his energies to restructuring the fourth-year undergraduate course in process design. He even taught this course on a part-time basis for three years after he retired in 1974. For one year after that, he worked for the dean as the College of Engineering liaison with the campus construction office. Finally, he retired completely, gave up editing the *Olin Hall News*, and cut his daily visits to Olin Hall to half days only. He continued his contacts with the alumni and his close association with the School until his death in 1986.

Throughout the thirty-two years of the Rhodes-Winding era, the primary concern of the School was undergraduate education. Dr. Winding fought hard to preserve the chemical engineering program during the establishment of the "common core" engineering curriculum and the change from a five-year to a four-year baccalaureate program. He was a steadying influence on the School through the turbulent years of the late 1960's. While he favored research and graduate work much more than Dusty Rhodes had, he did not agree with the College's infatuation with science to the detriment of engineering. "Engineering students should be educated for professional practice," he said, and to this end he devoted his career.

*Throughout the thirty-two years of the Rhodes-Winding era, the primary concern of the School was undergraduate education.*

# 8. The Bischoff Years

The five years under Director Kenneth B. Bischoff were the beginning of a long period of transition for the School. Ken was the first— to date, the only— director who had no previous Cornell affiliation. He was brought to Cornell to change things, especially research, in the School of Chemical Engineering. His own credentials were impressive: a 1961 Ph.D. from the Illinois Institute of Technology, followed by an NSF postdoctoral fellowship; faculty positions at the Universities of Texas and Maryland; several government-sponsored research projects. A consultant to the National Institutes of Health and the Esso Research Laboratories, he also served on committees for the Environmental Protection Agency and the Artificial Kidney–Chronic Uremia Program. He had chaired the first AIChE International Symposium on Chemical Reaction Engineering. His publications included more than fifty research papers and two books on chemical reaction engineering and bioengineering. He was thirty-four years old when he became director in 1970.

He soon hired three young research-minded assistant professors: John L. Anderson (Ph.D., University of Illinois) and James F. Stevenson (Ph.D., University of Wisconsin) in September 1972, and Michael L. Shuler (Ph.D., University of Minnesota), who had research interests in biochemical engineering, in January 1974. They all quickly obtained starter grants from NSF and other research support. By 1975 the research budget of the School was more than twice the level in the 1960's.

The number and quality of applicants for graduate work were unsatisfactory, however, especially to the younger faculty members. Nationwide, Cornell Chemical Engineering was not considered a leader in graduate studies or research, and the most promising B.S. graduates rarely applied. One problem was that money was tight; although the School received some $85,000 annually from industry, mostly for graduate-student support, it was clear that more fellowship money would be needed if the graduate program were to flourish.

The complexion of the faculty was also changed by several other events. In 1971 David Watt resigned to go to work at Procter & Gamble, and Jean Leinroth was denied tenure; he went to a faculty position at MIT and shortly thereafter to Crawford and Russell. In 1971–72 Vic Edwards spent a year at NSF in Washington D.C., and in 1973 he left Cornell to take a position with Merck. In 1972 Professor Wiegandt began working half time for Cornell, teaching during the fall term only; he spent the rest of the year as technical director for the Compagnie Francaise de Raffinage in Le Havre, France. Also in 1972 Professor Smith, after six years in the

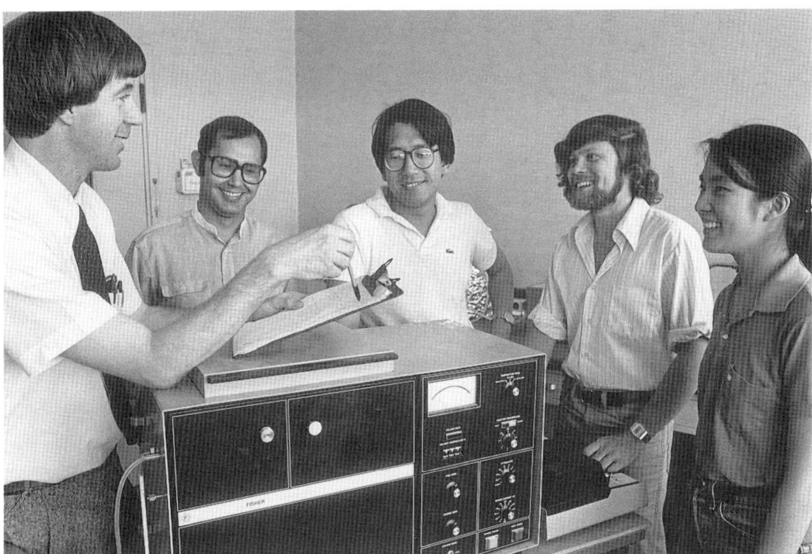

*Left: Professor Michael L. Shuler with students.*

*Left: Professor John L. Anderson in the classroom.*

*Left below: On the steps of Olin. By the 1970's the number of women in the School was beginning to increase.*

*James F. Stevenson*

*Dusty Rhodes' last visit to the School was in October of 1971, for the inauguration of the Fred H. Rhodes Professorship.*

*In the upper photograph, he is in the Fred H. Rhodes Lounge with Director Kenneth Bischoff (at center) and (top to bottom) Ted Buckenmaier '55, Bruce Davis '50, and Irwin Margiloff '52. Note the photograph of Rhodes on the wall.*

*The photograph below shows Rhodes with Davis and Buckenmaier in front of the mural in the lounge.*

dean's office, returned full time to Chemical Engineering. In 1973 he became associate director of the School. From 1972 to 1974, Dr. Robert E. Baier, head of chemical sciences at the Cornell Aeronautical Laboratory, served as the School's first adjunct professor; he helped Professor Anderson give a course called Applied Surface Chemistry.

In July 1975 Drs. Hedrick and Winding retired and were named emeritus professors. A retirement dinner had been given for Professor Winding on November 12, 1973, at the Union League in Philadelphia, during which it was announced that the Charles C. Winding Scholarship Fund had been set up for the support of Master of Engineering (Chemical) students.

A special dinner for Dusty Rhodes was held in Ithaca on October 21, 1971, to celebrate the establishment of the Fred H. Rhodes Professorship. This was the last time Dusty ever came to Ithaca. About one hundred alumni, plus spouses and faculty members and their spouses, attended the affair at the Warehouse Restaurant. The long fund drive had finally been successful; the professorship was formally given to Cornell with an endowment of $550,000 from over five hundred contributors.

A search committee consisting of Herbert D. Doan of Dow Chemical, Arthur Humphrey of the University of Pennsylvania, and James Wei of Mobil began looking for a prominent chemical engineer to fill the chair. Many were approached; several were made offers; none accepted. Finally, in 1974, Dean Edmund T. Cranch (who had succeeded Dean Schultz in 1972) agreed that the Rhodes professorship, with its requirement of much undergraduate teaching, should be filled from "inside" the existing chemical engineering faculty, provided that a search continued to fill a different chaired position with an "outside" distinguished professor. In April 1974, Dr. John R. Anderson of the University of Melbourne, Australia, an authority on chemisorption and catalysis, accepted an offer of the Herbert Fisk Johnson Professorship, which had been vacated when Dr. Winding retired. (Dr. Anderson later had family problems, however, and withdrew his name in 1975.) In July 1975, Professor Peter Harriott was named the Fred H. Rhodes Professor of Chemical Engineering, to Dusty's great satisfaction.

Director Bischoff was deeply interested in undergraduate teaching as well as in research, but his proposals for change were not always welcomed by the faculty. One idea that was enthusiastically accepted was a Research Honors program in which gifted students could enroll at the end of their third year; they could take graduate-level courses during their fourth year, and devote most of their time from then on to research and thesis preparation. By spending the summer in Ithaca after their senior year, these students could complete the requirements for the M.S. degree in one calendar year after receiving their BS. Four students entered this program in 1974 and similar numbers entered in subsequent years.

Less well received was the suggestion to rotate the teaching of required courses among the faculty members. Director Bischoff believed that, ideally, every faculty member should be able to teach every basic course, a view not shared by everyone. Some courses were rotated in 1973–74 and 1974–75, but not many; from this period comes the entry in the faculty minutes: "Professor Scheele offered to sell his soul if he didn't have to teach 5624. None of the faculty appeared interested in the purchase."

Even more controversial was the proposal to involve more faculty in the M.Eng. (Chemical) program, specifically in the process-design projects. The project sequence of two one-term courses, each carrying six credits, had been taught by Professor York since 1965. He did an excellent job, but effectively kept the other faculty members from participating in, or even learning much about, what he was doing. He was nearing retirement age, and someone else would soon have to take over. In April 1974 Professor Smith proposed splitting the fall-term course into two three-hour courses, one in process equipment design and one in economic analysis, and allowing the requirement for the spring-term design project to be either three or six credits. After lengthy, heated discussion, the motion was passed— with Director Bischoff breaking the tie vote. Professor York then announced that he would have nothing further to do with the M.Eng. program.

At a faculty meeting the next month, a professor moved to affirm that curriculum design and modification were the prerogative of the chemical engineering faculty, not the director, and that any substantial change in the content of a course— specifically, the undergraduate course in thermodynamics— be opposed. There was heated debate between Professor York and Professor Anderson, who was teaching the thermodynamics course at the time. A vote showed eight in favor of the motion, four against. Support for Director Bischoff was at all-time low.

Other curricular changes, on the other hand, were accepted in good spirit. In 1973 Professor Thorpe instituted a summer "audiovisual" course in Mass and Energy Balances, using slides or tapes the student could follow at his or her own pace. The College's Engineering Cooperative Program with industry was extended to Chemical Engineering in 1974; in this program, selected sophomores took their required fall-term courses in the summer, worked in industry from September through the winter break, and then followed their normal baccalaureate program with another stint in industry the following summer. Altogether, they spent about seven months working at industrial assignments, yet received their B.S. degree on time with the rest of their class. The co-op program quickly became popular among students and industrial firms. Because of scheduling problems in the summer, however, the organic chemistry requirement for co-op students had to be reduced from six to four credits. It was also necessary to offer two of the required third-year chemical engineering courses every summer.

The undergraduate curriculum was accredited for six years by Engineers' Council for Professional Development (ECPD) and AIChE in 1975. The accreditation inspector was Jack Tepe of Du Pont.

Student enrollments were steady during the Bischoff years. The number of B.S. degrees awarded annually ranged from thirty-four to fifty-five, averaging forty-three. Of these, one to three each year went to women students. In 1973 a woman, Jamie Sylvester, led the class academically with a 4.08 average.

For employment, 1971 was the worst of all years— four of the thirty-nine B.S. graduates had military obligations; twenty-four went to graduate school, many because they had no job offers; and all the others were unemployed at graduation time. The next year was somewhat better, but something new happened: seven of the graduates didn't bother to look for jobs or for anything else— like many others of their age, they needed time to "find themselves." Then, as the energy crisis of 1973–74 led to the formation of synthetic fuels programs across the nation, demand for chemical engineers picked up; in 1974 and 1975 there were more employment opportunities than people to fill them. By then most B.S. candidates knew what they wanted to do.

Graduate enrollments were also steady. Between ten and fifteen students enrolled each year in the M.Eng. (Chemical) program— fewer than hoped for, but more than was typical of the years to come. About eight M.S. and four Ph.D. degrees were awarded annually during this period. In the entire five years, only one woman graduated with an M.Eng. (Chemical) degree, one with an M.S., and none with a Ph.D.

Campus demonstrations continued into the 1970's. In an especially ugly incident in April 1972, students occupied Carpenter Hall for five days, demanding an end to ROTC and to the University's involvement with the Cornell Aeronautical Laboratory. A court injunction ended the occupation, which had been reasonably peaceful; in May, however, three hundred marchers broke windows in several campus buildings, including Olin Hall, and then continued their vandalism in Collegetown, where police ultimately used tear gas to disperse them. By 1973 such demonstrations had nearly disappeared, and through 1975 the campus was undisturbed by activists.

In 1973 Olin Hall was "invaded" by the College of Engineering when the Division of Basic Studies took over the north part of

*. . . as the energy crisis of 1973–74 led to the formation of synthetic fuels programs across the nation, demand for chemical engineers picked up. . . .*

the first floor. Room scheduling in the building was no longer the prerogative of Chemical Engineering; from 1971 on it was done by a College administrator. The chemical engineering library was moved to the second floor in 1973 and the position of librarian done away with; the library fell into disuse and the better books were moved to professors' offices for safekeeping. After years of neglect, the special-apparatus room had a brief resurgence of activity under a new attendant, but later relapsed into disarray. Olin Hall began to look its age.

In the spring of 1975, Dr. Bischoff decided not to seek a second five-year term as director of the School. To ease the transition to a new director, Dean Cranch appointed a search committee consisting of Professors Winding and Stevenson of Chemical Engineering and Professor Muetterties of Chemistry. Their recommendation was to appoint Associate Director Julian C. Smith, who was on leave at that time on a UNESCO project in Venezuela. His name was submitted to the chemical engineering faculty in March; it was approved, though far from unanimously. Smith became director on July 1, 1975.

Dr. Bischoff continued as professor in 1975–76, glad to get back to his research and outside activities. He had had a busy, eventful, stressful five years as director. During that time he had served a three-year term as a national director of AIChE and had been awarded, in 1973, the Ebert Prize of the Academy of Pharmaceutical Sciences for his paper on the distribution of anticancer drugs in the human body. At the School he had hired three faculty members, established the position of associate director, and strongly influenced the curricula and research efforts. He was young, voluble, and energetic, and worked hard to accomplish what he saw as his mission. There was no lack of communication: he seemed enamored of faculty meetings— four to six a month were common. But when he stepped down, he left behind a contentious faculty, divided, almost polarized, in their opinions on many issues. In June 1976, Dr. Bischoff resigned to take a faculty position at the University of Delaware, where within a few years he became chairman of the Department of Chemical Engineering.

# 9. Director Smith and the Growth of Research

Julian C. Smith was director of the School for eight years, from 1975 to 1983. His first year was a quiet one, with no major changes, in order to give the faculty differences a chance to be resolved. Smith said his most audacious act that year was to move the box for U.S. mail out of the business office into the hallway.

Teaching went on as before, except that classes were large, with some seventy-five students in the third year. Professor Smith began offering a two-hour course in Industrial Organic Chemistry to the third-year co-op students; Professor Scheele started using Monsanto's "Flowtran" program, by long-distance telephone, for his elective course in computer-aided design. Francis Sherman was named the first Winding Scholar. The AIChE student chapter, advised and inspired by Professor Stevenson, won a national "Outstanding Chapter" award. Geological Sciences was given Room B-49 to use until a new building was available. The only jarring notes were the resignation of Professor Anderson to become an associate professor at Carnegie-Mellon University, and a 1960's-style campus demonstration sparked by the firing of a Black administrator in Cornell's minority program.

Things began to happen in 1976–77. During the previous year, a search had been going on to fill the Johnson professorship and, for the first time in the School's history, advertisements for the position were placed in professional journals. These were supplemented by the usual letters to chemical engineering department heads and industrial leaders. The efforts paid off: fifty-five applications were received, and seven candidates were brought to Cornell to give seminars and meet the faculty. Two of them were so impresssive that Dean Cranch authorized Director Smith to offer a chaired professorship to each of them. Both accepted. In September 1976, Keith E. Gubbins from the University of Florida came to Cornell as the Thomas R. Briggs Professor of Engineering[1]; in the spring of 1977, Robert P. Merrill of the University of California, Berkeley, became the Herbert Fisk Johnson Professor of Industrial Chemistry. Both were distinguished leaders in their respective fields of research. Their arrival marked a turning point for the School.

These appointments, unlike earlier ones, required substantial financial commitments by the School and the College to provide the needed research facilities.

---

[1]Thomas R. Briggs was a highly respected chemistry professor who taught physical chemistry to chemical engineering students in the 1930's and 1940's. His daughter Elizabeth Adelaide Briggs painted the mural in the Olin Hall Lounge in 1949.

*George Scheele*

46


*Right: Francis Sherman '76 (at left) was the first Charles C. Winding Scholar. With him are Professors Winding and Smith.*

*Below: Professors Gubbins (in the photo at left) and Merrill (in the photo at right) with students.*

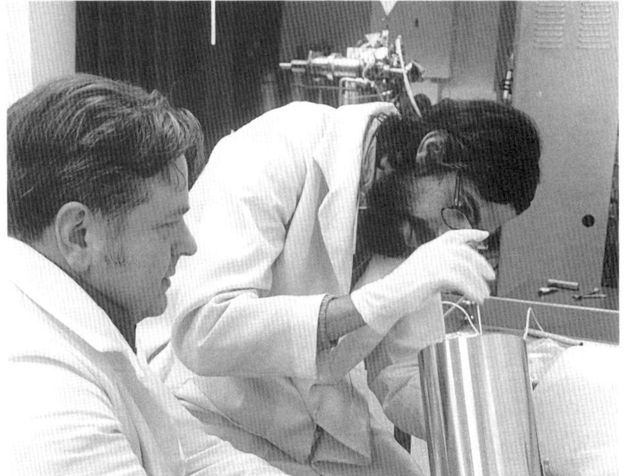

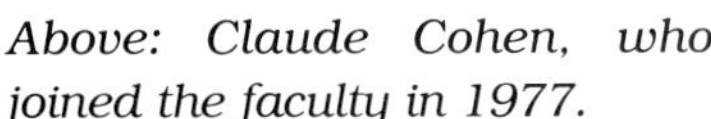

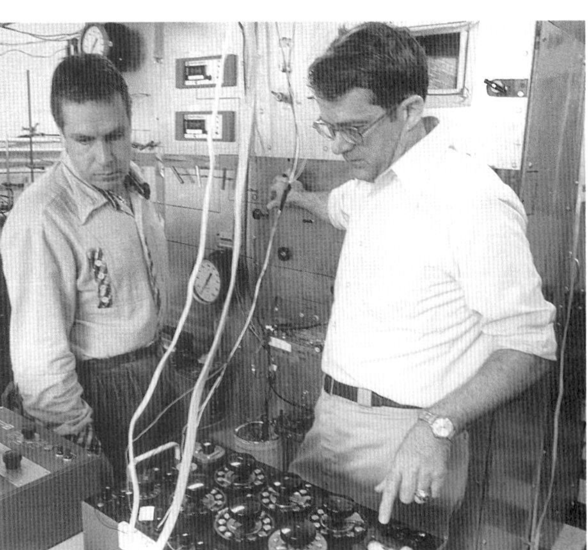

*Above: Claude Cohen, who joined the faculty in 1977.*

*Above: Adjunct Professor Jorge Calado with William B. Streett.*

Professor Merrill's chief requirement was a place to house his ultrahigh-vacuum equipment for studying surface phenomena such as chemisorption and catalysis. While at Berkeley, he had created, with U.S. Air Force sponsorship, an extensive laboratory for this purpose. The University of California had never taken title to the equipment; consequently, it was possible to move the laboratory units, valued at $500,000, for a cost of about $37,000. Providing space for them was another matter: a "clean room," with elaborate air filters and conditioners and about 3,600 square feet of floor space, had to be constructed out of a large basement room formerly used for equipment storage. For this, Dean Cranch provided the required $450,000 from College funds.

Professor Gubbins had different needs. His field was thermodynamics and statistical mechanics, with emphasis on study of the properties of atoms and molecules through computer simulation. His computing needs were greater than could be supplied by the University's IBM 370/168, but money was not available to provide additional computing capability. The situation appeared dark until, through a fortunate circumstance, funds appeared from an unexpected source. Donald F. Berth, the College's director of special projects, had approached the Joseph N. Pew Foundation with a $10-million request for a College of Engineering project; as an afterthought, an additional $500,000 was added to the proposal "for the renovation of Olin Hall." The afterthought was funded; the main project was not. Using part of this grant, obtained with no effort on his part, Director Smith was able to make a computer facility out of two first-floor rooms and pay half the cost of a Digital PDP 11/70 minicomputer. (The rest of the cost was paid for by a grant from the NSF.) This was the first departmental computing facility in the College of Engineering.

In July 1976, Professor Bischoff unexpectedly resigned to go to the University of Delaware, and the next year Jim Stevenson, after being promoted to associate professor, left for General Tire and Rubber in Akron, Ohio. All the faculty members hired between 1965 and 1972 had, for various reasons, left the School.

In 1975–76 a search had begun for a new assistant professor. Several offers were made that year, but the competition was stiff and all the candidates accepted positions elsewhere. In 1976–77 the recruiting efforts were intensified, and after a two-year search involving one hundred eleven applicants and seventeen campus interviews, the School hired Claude Cohen, who had a Ph.D. in physical chemistry from Princeton. He was then a postdoctoral fellow in chemical engineering at CalTech. A new faculty search was made in 1977–78 without success; however, a retired Army colonel, William B. Streett, was appointed as a senior research associate. Dr. Streett brought with him from West Point some high-precision equipment for measuring thermodynamic

*. . . as an afterthought, an additional $500,000 was added to the proposal . . . . The afterthought was funded; the main project was not.*

48

*Above: Professor Merrill's labora-tory, "before" and "after".*

*Below: The VAX computer room, "before" and "after".*

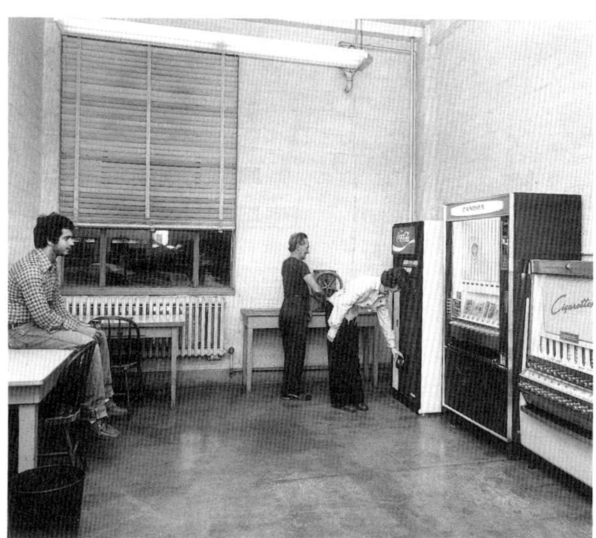

*Paul Steen*

*William Olbricht and Joseph Cocchetto*

properties at elevated pressures. In January 1978, six months before his scheduled retirement, Professor York died of a heart attack. Two assistant professors were added in the fall of 1979: William Olbricht and Joseph Cocchetto, who had doctoral degrees from CalTech and MIT, respectively.

The first four years under Director Smith saw great changes in the faculty, the teaching loads, and the facilities. In addition to the "clean room" and the computer facility, the School acquired its own truck, and new construction provided a micro-biology laboratory on the third floor, nine refurbished laboratories in the east wing, an expanded electrical system, and a new roof. The Unit Operations Laboratory was painted for the first time in thirty-eight years. (Most of these changes were funded by the Pew Foundation grant.) Undergraduate classes swelled to record levels. The pace of research rose at a phenomenal rate: in 1979 the expenditures of $600,000 were three times the figure for 1975. The one area of weakness was in the recruitment of graduate students, for the number and especially the quality of M.S./Ph.D. applicants were clearly inadequate to meet the needs of the newer faculty members. In 1976 Director Smith established a standing committee for graduate-student recruitment, with Professor Shuler as chairman; by 1979 the committee's efforts had borne fruit. Some of the very best graduate student candidates, domestic and foreign, now applied to the School. Chemical Engineering at Cornell was becoming known as a leading research department.

During the next four years, the changes continued, though at a slower pace. Dean Cranch left in 1978 to become president of Worcester Polytechnic Institute, and Thomas E. Everhart came from Berkeley in 1979 to take his place. Dean Everhart, fortunately, continued to support the developments in Chemical Engineering. In 1980 Director Smith was reappointed for a second term. In 1981 Dr. Streett was appointed to the rank of professor with tenure and became a full member of the Chemical Engineering faculty; soon afterward, however, he accepted a half-time position as associate dean of the College. Also in 1981, Professor Cocks took early retirement, for in 1977 the laboratory course in his specialty, chemical microscopy, had been dropped from the chemical engineering curriculum and prospects for increased research in this area were unpromising. Director Smith recommended that Professor Cocks accept a position offered by the Los Alamos National Laboratory, which he did. He was named professor emeritus. After an extensive search, Paul Steen, a Ph.D. from Johns Hopkins, was appointed an assistant professor in 1982, as was Douglas Clark, a Ph.D. from CalTech, in 1983. Also in 1983, Dr. Jorge Calado, a thermodynamicist from the University of Lisbon, Portugal, was named an adjunct professor.

This period saw the first visiting professors hired from

*Douglas Clark*

50



*Right: Dean Thomas E. Everhart (left) and Director Smith with Ellie Moulton at the party held in her honor when she retired after serving as administrative secretary to three directors of the School.*

*Right: The Fred H. Rhodes Lounge in Olin Hall.*

*Below: Marjorie Leigh Hart (B.Ch.E. '51), a member of the Chemical Engineering Advisory Council for many years. She is now on the Engineering College Council.*

industry to teach required undergraduate courses. They were all Cornell alumni. Ken Karmel (B.Ch.E. '54) of the Ethyl Corporation and Joe Degenfelder (B.Ch.E. '61) of Catalytic, Inc. taught the senior design course in 1982 and 1983, respectively, with support by a grant from Chevron USA. Fred Vorhis (B.Ch.E. '66, M.S. '68) of Chevron ran Unit Operations Laboratory in the fall of 1982.

In 1979 Eleanor Moulton retired; she had been administrative secretary to Directors Winding, Bischoff, and Smith since 1957. June Williams was hired and held the position until late 1982. Smith, who found himself increasingly involved in routine day-to-day matters of finance and personnel, attempted to find an administrative manager for the School, but without success.

More successful was his establishment of a Chemical Engineering Advisory Council. This had been suggested, repeatedly, by Donald Berth; it was finally implemented in 1980–81. Twelve members were appointed— ten from industry and two from academia. All but three were alumni of the School. The Council first met in Ithaca in April 1981, under the chairmanship of Jim Donnalley of General Electric, to discuss the balance between teaching and research, faculty and staff needs, and upgrading of facilities. It convened in both fall and spring in subsequent years, and was a source of good ideas and support. The membership of the Council is shown in Appendix F.

More changes were made in Olin Hall. Rooms on the third floor were refitted as laboratories for studies of rheology and thermodynamic properties; the Process Control Laboratory was equipped through a grant from the Sun Company; Chevron USA funded a Kinetics and Catalysis Laboratory. Additions were made to the computing facility, and a Hewlett-Packard data-acquisition system was installed in the Unit Operations Laboratory. A new lathe and a milling machine were added to the machine shop. In 1980 Joseph Coors (Ch.E. '40) provided funds for renovating the student lounge, which was renamed for Fred H. Rhodes. The mural, an allegory of the five-year chemical engineering curriculum, was restored by David Finn, son of Professor Robert Finn.

Research expenditures reached almost $1.3 million in 1982–83, over six times the 1975 figure. The fifty-three active sponsored projects testified to the interest and ability of the faculty to obtain outside support. Among them were eleven projects which resulted from a $250,000 research initiation grant made in 1982 by the Sun Company to support the faculty in trying out promising ideas not sufficiently well demonstrated to attract funding from the usual sponsoring agencies. The research efforts were now in four major areas: molecular thermodynamics; catalysis and surface science; polymers, fluid mechanics and rheology; and biochemical engineering. The last of these was given a boost in 1982 by the creation of a Biotechnology Institute at Cornell, sponsored by New York State and three industrial contributors; Professors Finn, Shuler, and Clark soon became active members.

The undergraduate curriculum shown in Appendix C for 1977–78 differs only a little from the one ten years earlier. Constraints imposed on the four-year program by the College and by accrediting agencies left little room for change. Only one term of physics was required in the first year, instead of two; the space was filled by a course in "freshman engineering." Engineering drawing disappeared. The course in microscopy, so long a part of the curriculum, was deleted, as were electrical engineering and the project laboratory. The biggest change was the addition of a process design course, Chemical Process Synthesis, in the fourth year; this was done in 1971 so that the B.S. program could meet ECPD accreditation requirements. Total required credits were 136 for the B.S. degree and 166 for the M.Eng. (Chemical). The undergraduate curriculum was accredited for six years by ECPD in 1975 and by the Accreditation Board for Engineering and Technology (ABET), successor to ECPD, in 1981.

Undergraduate enrollments continued to be large: an all-time record of eighty-one B.S. degrees were awarded in 1983. Women now made up one-quarter to one-third of the students in each class. Jobs were plentiful until the spring of 1982, when

*Women now made up one-quarter to one-third of the students in each class.*

suddenly most companies stopped hiring chemical engineers. Freshmen immediately began choosing other engineering disciplines, but four years had to elapse before this was reflected in the number of degrees awarded. Graduate-student enrollments rose a little. The number in the M.Eng. (Chemical) program was steady at about eight; M.S./Ph.D. candidates increased from forty to over fifty, with a strong trend away from the terminal M.S. degree so popular in earlier years. By 1983 nearly all the entering M.S./Ph.D. students planned to complete a doctorate. Space in Olin Hall was in very short supply.

More Ph.D.'s meant that more support money was needed. Some came from the dean's office, some from new industrial fellowships. Industrial contributions, in the period of great demand for chemical engineers, increased rapidly. Grants for new young faculty members, to help them get established, were given for several years by Du Pont. Responding to the pressing industrial need for more advanced graduates with United States citizenship, several firms created generous fellowships specifically for American Ph.D. candidates.

In the spring of 1983, Professor Smith took a sabbatical leave and in June stepped down from the directorship. Professor Rodriguez was acting director during the spring term; Professor Gubbins became the new director on July 1. By then the School was very different from what it had been eight years earlier, when Director Smith took office. Six professors had left and seven had been added. Research had increased more than six-fold; new computing facilities and some sophisticated laboratories and equipment had been installed. Dissension in the faculty had largely disappeared. The number and quality of graduate students had improved, and the School's reputation for research had risen considerably: in the 1982 Jones-Lindzey report *Assessment of Research Doctorate Programs in the United States*, Cornell Chemical Engineering was ranked sixteenth nationwide and rated as one of the most rapidly improving departments.

On the negative side, some serious problems awaited the new director. Undergraduate enrollments, though still high, were expected to drop back to or below pre-1976 levels. Several professors were nearing retirement and replacements would have to be found. More and better administrative personnel and facilities were urgent needs. Money was tight: as the demand for chemical engineers fell, contributions from industry had diminished, and for a number of reasons the College of Engineering was no longer able to cover operating deficits. Despite the recent improvements, Olin Hall was clearly unsuited to the School's future needs in research and graduate instruction; the only remedy would be a major renovation of the entire building.

It was a time of both problems and promise.

*[In 1982] Cornell Chemical Engineering was ranked sixteenth nationwide and rated as one of the most rapidly improving departments.*

# 10. Director Gubbins
# and the School of Today

"The concentration of men retiring in 1981–1989 is unfortunate," wrote Dr. Rhodes in 1957, after analyzing the age distribution of his faculty, "but it resulted from the rapid growth of the School of Chemical Engineering in the early post-war years. Much younger men were not qualified, and older ones were not available." And now, in 1988, the "concentration" of retirements, delayed a little by the increase in retirement age from sixty-five to seventy, has arrived. What is amazing is that all five of the chemical engineering professors on Dusty Rhodes' list are still affiliated with Cornell. Three of them are still teaching.

Keith E. Gubbins, the fifth director of the School, has responded with vigor to the problem of replacing approximately one-third of his faculty. In the four years since he took office, there have been many faculty changes, including and in addition to the expected retirements. Professors Smith and Wiegandt retired in 1986 and 1987, respectively; Professor Streett became dean of the College of Engineering, full-time, in 1985, replacing Dean Everhart; Professor Cocchetto left Cornell in 1984. Douglas Clark received one of the prestigious and lucrative Presidential Young Investigator awards, but soon after, in August 1985, he resigned to join the faculty of the University of California at Berkeley. In 1987 Professor Thorpe went on "phased retirement" and now teaches only during the fall term.

*Keith Gubbins*

To offset these losses, there has been a continuous search for new faculty— with gratifying results. The School is now competitive with other top chemical engineering departments nationwide for the very best faculty candidates. Recent additions of assistant professors, all of great promise in research and teaching, are: Paulette Clancy (University of Oxford), appointed in 1983; Brad Anton (CalTech), Peter Clark (CMU and Westvaco Corporation), Don Koch (MIT), and Athanassios Panagiotopoulis (MIT), all of whom started in 1987; and Dan Hammer (University of Pennsylvania), who joined the faculty in 1988. Also added was John Zollweg, formerly a research associate, who was named an associate professor, without tenure, in July 1986.

*Herbert Wiegandt*

These additions, however, did not satisfy all the needs for teaching certain undergraduate courses. To fill these gaps, Helen Downs Haller has been appointed as a lecturer to run the Unit Operations Laboratory and help with the senior process design course. Haller got her B.Ch.E. at Cornell in 1964 and was the School's first woman Ph.D. in 1967. Also helping is Clarence Shoch, B.Ch.E. '51 and recently retired from Du Pont, who for two years has been a visiting professor teaching process design; in

*Paulette Clancy*

*Brad Anton*

*Peter Clark*

*Donald Koch*

addition, he has supervised several Master of Engineering design projects.

Other faculty-related events during the four-year period included two gala dinners for Professor Smith, one in 1983 to honor him at the end of his directorship, and the other after his retirement in 1986. The Julian C. Smith Lectureship, funded by gifts from alumni, was established (Professor R. Byron Bird of the University of Wisconsin was the first Smith lecturer in March 1988). National awards were won by three professors: Robert Finn received the 1985 Food, Pharmaceutical and Bioengineering Award of AIChE; Michael Shuler was given the 1986 Marvin J. Johnson Award from the Microbial and Biochemical Technology Division of the American Chemical Society; and Keith Gubbins won the 1986 AIChE Alpha Chi Sigma Award. Professor Finn was also honored by his colleagues in Julich, Germany, who named a newly isolated bacterium *Thermoanaerobacter finnii.* Dr. Margarida Telo da Gama of the University of Lisbon, Portugal, a frequent visitor in the thermodynamics research group, was appointed an adjunct professor in 1986. In 1986–87 Professor Gubbins was on leave doing research at the University of Oxford, while Professor Shuler served as acting director; in the fall of 1987, Gubbins returned to begin a second three-year term as director.

The School lost two former faculty members with the deaths of Professor Mason in December 1983, and of former Director Winding in March 1986.

Research has continued to flourish under Director Gubbins: expenditures are now about $2.5 million annually. Research has become collaborative and interdisciplinary through joint projects with Chemistry, Applied Physics, Mechanical and Aerospace Engineering, Food Science, and other academic units, and through involvement in the Materials Science Center, the Cornell High Energy Synchrotron Source (CHESS), the National Nanofabrication Facility, the Biotechnology Institute, and the Center for Theory and Simulation in Science and Engineering (a national supercomputer facility, sometimes referred to as the Theory Center). Graduate students from Chemistry, Physics, and Materials Science and Engineering have done their thesis studies under the direction of chemical engineering professors. Today, research is a major activity of the School. It's also a big business.

In 1983 Director Gubbins hired an administrative manager—Jeffrey Curtis—to manage the School's financial and administrative affairs. Curtis brought badly needed order and control into the financial system and relieved the director of many routine matters. In 1985, when he went to another managerial position at Cornell, his place was taken by Anne K. Scott, who was replaced in 1986 by Betty J. Bortz. Meanwhile, other new staff positions had been established. The position of executive assistant was filled successively by Demetra Dentes, Jeanne Mosher (Yarussi), and Carol Brewer. Jan Haldeman was appointed communications specialist. Later she was replaced by Nancy Morris as coordinator for external relations. All of these staff members who left Chemical Engineering moved on to better and more responsible positions. (Jeanne Mosher, for example, left to become a wife, stepmother, and homemaker). The most recent appointment, in 1986, was that of Margaret Murray as accounts coordinator. (In January 1987 she received the Dedicated Service Award given to outstanding University employees.)

The present (1987–88) undergraduate curriculum, shown in Appendix C, differs from the 1977–78 curriculum in several ways. Among the changes are the following: chemical engineering thermodynamics is taught in one term instead of two; the subjects of fluid mechanics and of heat and mass transfer are covered in two separate courses; process control is now required; and the chemical processes course has been replaced by a chemical engineering elective. Six elective courses must be chosen from three areas: humanities or history; social sciences; and expressive or language arts. The changes seem pretty small, however, compared with the wholesale alterations common in the early days of the School. The total number of credits required for the B.S. degree is 137. The B.S. program was accredited by ABET for six years in 1987.

In 1985 the College of Engineering initiated a two-year Master of Engineering cooperative program in which students spend a year gainfully employed in industry. To date, three chemical engineering students have availed themselves of this opportunity. In accordance with advice from the Advisory Council, the M.Eng. (Chemical) program was changed significantly in 1987 and the course requirements made more like those in the graduate research programs. M.Eng. (Chemical) students are now required to choose two courses from each of two groups: Group I consists of Computer Aided Design, Equipment Design, and Reactor Design; Group II comprises the advanced courses Reaction Kinetics, Thermodynamics, Transport Phenomena, and Applied Mathematics. Three of the Group II courses are required of M.S. candidates; all four of them must be taken by candidates for the Ph.D.

Undergraduate enrollments were large through 1984–85, then fell sharply. Only thirty-four B.S. degrees were awarded in 1986 and thirty-three in 1987 (Appendix E). This reflected a nationwide trend as the 1982 drop in demand for chemical engineers had its effect, and as the public image of chemists and chemical engineers, in the aftermath of Bhopal and similar disasters, became badly tarnished. No future increases— in fact, possible further decreases— in the number of undergraduate chemical engineering students are likely for the next several years. The

*Athanassios Panagiotopoulos*

*Daniel Hammer*

*John Zollweg*

*Margarida Telo da Gama*

56



*Right: Olin Hall in the fall of 1987. Renovation of the east wing is in progress.*

*Right: One of three chaired professors now on the School faculty is Peter Harriott, the Fred H. Rhodes Professor. (The others are Keith E. Gubbins and Robert P. Merrill.)*

*Below: Margaret Murray, the School's accounts coordinator, recently received the Dedicated Service Award at Cornell.*

*Right: Chemical engineering senior Geoffrey C. Achilles was chosen as one of Cornell's 1987 Presidential Scholars. He named Professor Raymond G. Thorpe as his most influential teacher. Thorpe is the only Cornell professor to have been selected by a Presidential Scholar every year since the program began in 1984.*

population of graduate students has continued to grow, though slowly. M.Eng. (Chemical) degrees numbered ten to twelve in 1984 through 1987, much the same as in earlier years. The number of M.S. degrees has decreased as more students choose to go directly for their Ph.D.; the number of Ph.D. degrees has begun to rise. In 1987 ten Ph.D. degrees, a record number for one year, were awarded.

Women now make up almost one-third of the undergraduate students and about 20 percent of the graduate students.

The Chemical Engineering Advisory Council expanded to eighteen members as Keith Gubbins added prominent individuals from industry and other universities (see page 61 and Appendix F). The group met twice annually from its inception until 1986, when it decided that henceforth once a year would be often enough. The Council has given the School a great deal of support and guidance through the recent period of change and renovation.

One of Director Gubbins' principal goals, the renovation of Olin Hall on a grand scale, is becoming a reality. In 1983 the University funded an architectural study of the entire building. The recommendations included grouping faculty offices and reducing their size; creating new laboratories suitable for advanced research; improving the lecture rooms; and making general improvements to meet current building codes. The steel-frame building, despite its forty-five years, is basically sound, and the interior walls, none of which is load-bearing, can be moved as needed. When the Theory Center was given a temporary home in Olin Hall in 1985, it appeared for a time that money would be made available to renovate a number of offices in the north wing, but the next year the funds evaporated and the deal fell through.

About the same time, a fund drive was initiated by the College's development office to raise $5–7 million to renovate the east wing of the building. After a slow start, this drive has been successful: Eastman Kodak, Harry Mattin (B.Chem. '18), and Mattin's company, the Mearl Corporation, have contributed $1 million each, and many other gifts, large and small, from alumni and industrial firms, have swelled the fund to over $5 million. Construction started in August 1987. Twenty-three new offices and laboratories, many equipped for biochemical engineering research, are being provided, in part by flooring over space in the "high bay" of the Unit Operations Laboratory. This empty space, originally designed to house tall columns and other large chemical engineering equipment, has never been used for that purpose. Olin Hall will also get a new stairway and its first passenger elevator.

Director Gubbins has steadily upgraded the computer facilities of the School. For graduate research, the PDP 11/70 minicomputer was replaced in 1984 with a more powerful



*One of Director Gubbins' principal goals, the renovation of Olin Hall on a grand scale, is becoming a reality.*

*Right: Director Gubbins and Professor Smith look over the Unit Operations Laboratory space that is being converted to new offices and laboratories.*

*Below: Presentation by alumni of the class of 1960 of funds for the Instructional Computing Facility. Director Gubbins holds the plaque that was later mounted in the laboratory.*

VAX 750; in 1987 this was further upgraded to a VAX 8250. The special apparatus room was cleaned out and converted to an air-conditioned terminal room. Dr. Steven Thompson, a senior research associate, is in charge of this facility. A new University telephone system made it possible to create computer networks inside Olin Hall and with other schools and departments.

These new computer facilities, good as they were, did little to help the undergraduates, who still relied heavily on the overburdened central computer system of the University. Limited access to the terminals often made it difficult for them to use available programs in their courses, especially in process design, or to develop their own programming skills. Hearing of this, the Chemical Engineering Class of 1960 undertook, as a twenty-five-year reunion project, to raise funds for an Instructional Computing Facility (ICF) to be used exclusively by undergraduates. It would have several microcomputers and printers and connections to the VAX computer for lengthy calculations. Led by Ken Ackley, Joe Degenfelder, and George Roberts, the class set out to raise $50,000. It received well over $85,000, and presented the much-appreciated gift to the School in June 1985. It took time to fund and construct a suitable home for the equipment in Room B-49— vacant after Geological Sciences moved to Snee Hall— and former storage room B-53, but the job was done at last. The completed ICF was dedicated with a ceremony in May 1987.

This brings things pretty well up to the present day. What does the future hold for the School? For the next few years, some things are reasonably clear. Undergraduate enrollments will be low and research and graduate studies will expand. A Golden Jubilee celebrating the School's first fifty years is being held in 1988. Three faculty retirements are in the offing: Professor Von Berg in 1988, Professor Finn in 1990, and Professor Thorpe in 1991. One or two assistant professors will probably be added to the faculty, and possibly a more senior professor as well. Renovation of the lower floors of the east wing of Olin Hall should be complete by December 1988. A Building Fund Committee of prominent alumni, chaired by Herbert D. Doan (B.Ch.E. '49), is currently soliciting all alumni of the School for contributions to the renovation fund; if this drive is successful, the third floor of the east wing will be renovated during the current construction project. The Theory Center is expected to move out of Olin Hall as new space becomes available elsewhere.

All this should take place within the next three years or so; for times beyond that, the crystal ball is clouded. Some speculations as to longer-term developments in the School of Chemical Engineering, and a look at the sweeping changes that have occurred in the School since its founding in 1938, are the subjects of the next (and last) chapter of this history.

*. . . the class set out to raise $50,000. It received well over $85,000. . . .*

# 11. The Past, the Present, the Future

In fifty years, things change. Looking back at my freshman year of 1937–38, I see little that has stayed the same. Many of the changes were small and almost imperceptible at the time; others were large and often painful. Taken together, over the years, they have transformed the School of Chemical Engineering in nearly all its aspects: its faculty, staff, students, curricula, research, facilities— and its relation to the College of Engineering and the rest of the University.

The faculty in 1937–38 consisted of two professors, Rhodes and Winding. Both of them had Ph.D.'s, unusual at a time when only 7 percent of the College faculty had doctoral degrees. Neither had tenure, for tenure wasn't established at Cornell until 1939 and not legislated in its present form by the trustees until 1948. The 1987–88 faculty roster, by contrast, lists the names of twenty chemical engineering professors of various ranks, all but one of whom have tenure or are on a tenure track. Only one does not have a doctoral degree. One is dean of the College of Engineering, full-time, and one is on phased retirement, teaching half-time. In addition, there are two adjunct professors, one lecturer, one senior research associate, and two visiting professors, for a grand total of twenty-six.

Since 1938, thirty-six individuals have been added to the professorial ranks of the School (not including adjunct or visiting professors). Nineteen have left the faculty: seven retired (four of whom have since died), one died before retirement, three were denied tenure, and eight left the School for various other reasons. These changes are shown graphically in Appendix A-3.

The support staff has grown, too. In 1938 it consisted of one secretary and one mechanician, Harold C. Scott. Total: two. ("Scotty" moved with the School to Olin Hall in 1943 and worked there until his death in 1952.) In 1988 the staff includes an administrative manager, an executive staff assistant, an accounts coordinator, a coordinator of external relations, four administrative aides, two secretaries, one research support specialist, and two laboratory equipment technicians. Total: thirteen.

By June 1987 the School had awarded 2,207 undergraduate degrees (93 Ch.E., 65 B.S.Ch.E., 894 B.Ch.E., and 1,155 B.S.) and 735 graduate degrees [45 M.Ch.E., 294 M.Eng. (Chemical), 228 M.S., and 168 Ph.D.]. Its living alumni number well over two thousand, and throughout the history of the School, its graduates have been extraordinarily generous, supportive, and loyal. Many have given lectures to fourth-year students or AIChE groups; some have helped teach undergraduate and graduate-level design courses; others have organized and contributed to special fund drives, or served on the Chemical Engineering Advisory Council. The present success of the School is due in no small measure to the efforts and contributions of its alumni.

In many ways, the chemical engineering students haven't changed much in fifty years—they're still young, bright, and energetic. They are inexperienced, though not quite as much as we were. Now almost a third of them are women, however, compared with none fifty years ago. There were no Black students then, either, and few other minority or foreign students. In 1938 fewer students were children of professional people and more were from families of shopkeepers and blue-collar workers; today many are children of engineers and scientists and lawyers and doctors. In recent years, more than a few have been children of our own alumni. Today it costs more than $17,000 a year to study engineering at Cornell. Tuition and fees, fifty years ago, were $400 a year and room and board were $500 or less; even so, a great many people could not afford the costs at Cornell or at any university, for that matter.

The size of the student population has changed, especially at the graduate level. The numbers of degrees awarded are revealing: a comparison of the figures for 1937 and for 1987 are given in the table on page 63.

*Below: The Chemical Engineering Advisory Council at the October 1987 meeting. (Cornell degrees are indicated.)*

*Top row, left to right: H. Ted Davis, University of Minnesota; David S. Barmby, Sun Company; Scott C. Roberts, Shell Oil; L. Gary Leal, California Institute of Technology; Thomas K. Smith '62, Dow Chemical; Vern W. Weekman, Mobil Oil; and George W. Roberts '61, Air Products and Chemicals.*

*Bottom row, left to right: Stanford H. Taylor '51, Aero-Vironment; Bryce I. MacDonald '45, General Electric; John F. Schmutz '55, Du Pont; David S. Laity, Chevron Research; William R. Schowalter, Princeton University; Leonard A. Barnstone Ph.D. '65, Exxon Research and Engineering; and Andreas Acrivos, Stanford University.*

*Missing are Samuel C. Fleming '63, Arthur D. Little, Inc.; W. Thomas Mitchell, Biochemicals Business; and James Wei, MIT.*

Above: The class of 1964.

Right: The class of 1977.

Degrees Awarded Fifty Years Ago and Now

|                        | 1937 | 1987 |
| ---------------------- | ---- | ---- |
| Baccalaureate degrees  | 7*   | 33   |
| Advanced degrees       | 1    | 27   |

*These "Chemical Engineer" degrees were technically postgraduate degrees, since their recipients already had the four-year Bachelor of Chemistry degree; however, the Ch.E. degree was administered by the College of Engineering, not the Graduate School.*

In 1938 there were about 234 undergraduates in the School (roughly 110 freshmen, 60 sophomores, 30 juniors, 20 seniors, and 14 fifth-year students, known as "hangovers"), and two or three graduate students. Now there are no freshmen or sophomores (they are enrolled in the College), 29 juniors, 33 seniors, and 74 graduate students (eight M.Eng. and 68 M.S./Ph.D. ). As discussed earlier, the numbers of upperclass undergraduates fluctuated wildly during World War II and immediately thereafter, stayed fairly constant from 1953 to 1975, rose to record heights between 1976 and 1985, and then fell sharply to the present level. The rise in graduate-student population has been steady over the years, except for the surge between 1966 through 1969, when fifth-year students became M.Eng. candidates.

Before the 1960's, students did a lot more singing than they do at present. They sang at Zinck's and the Dutch Kitchen and other local watering places; at fraternity events; at informal gatherings; at Friday night beer parties in the Olin Hall lounge. They knew all the words—most of them, anyway—to many songs. Dusty Rhodes, as an *aide memoire* for his students, had printed a booklet entitled *Ye Songs—Both Sprightlie and Doleful—of Ye Honourable and Dystinguyshed Companie of Ye Chymical Engineers.* The songs and singing are largely gone. Who today sings "Silver Dollar" or "In Bohemia Hall" or "Landlord, Fill the Flowing Bowl"?

Were the undergraduate students in the early days of the School smarter than they are now? I don't think so; I believe today's students are at least as gifted as their predecessors. Matriculants were less well prepared, years ago, than they are now, for few came with any knowledge of calculus, and advanced high-school courses in physics and chemistry were things of the future. Students did work harder and longer for their baccalaureate degrees—five years instead of four, and terms of fifteen weeks instead of thirteen or so. Eight-o'clock classes were much more common and Saturday mornings were often completely taken up by five-hour chemistry laboratory periods. Calculations were long and tedious, for log-log duplex slide rules are orders of magnitude

*The songs and singing are largely gone. Who today sings "Silver Dollar" or "In Bohemia Hall" or "Landlord, Fill the Flowing Bowl"?*

slower than a microcomputer or even a pocket calculator.[1] The students complained bitterly about their work load, as do students today; one of the few constants in the sea of change is the reaction of twenty-year-olds to pedagogical demands on their time. There is another constant also, a most important one: the pride the graduates take in having surmounted the obstacles and survived the rigors of the "impossible" Chemical Engineering program.

Appendix C shows the undergraduate (B.Ch.E.) curricula for 1937–38, 1947–48 and 1957–58; for the succeeding ten-year periods it shows the B.S. and the M.Eng. (Chemical) curricula. Overall requirements, by category, are summarized for 1937–38 and 1987–88 in the table below.

The noteworthy changes over the fifty-year period are as follows: The chemistry requirement has been cut by over 50 percent, with the elimination of analytical chemistry. The extensive requirements in mechanical and electrical engineering have

---

[1] A few students had pocket calculators in the late 1960's, but weren't allowed to use them in examinations until about 1971. Then, within one year, slide rules virtually disappeared.

## Degree Requirements Fifty Years Ago and Now

| Subject Area | 1937–38 Five Years | | 1987–88 Four Years | | Five Years | |
|---|---|---|---|---|---|---|
| | No. of Courses* | Credit Hours | No. of Courses | Credit Hours | No. of Courses | Credit Hours |
| Chemistry | 21 | 60 | 9 | 26 | 9 | 26 |
| Mathematics | 2 | 10 | 4 | 16 | 4 | 16 |
| Physics | 4 | 14 | 3 | 12 | 3 | 12 |
| English (writing) | 2 | 6 | 2 | 6 | 2 | 6 |
| Chemical Engineering | 6 | 16 | 10 | 34 | 15 | 49 |
| Mechanical Engineering | 15 | 40 | – | – | – | – |
| Electrical Engineering | 2 | 8 | – | – | – | – |
| Economics | 1 | 3 | – | – | – | – |
| Mineralogy | 1 | 3 | – | – | – | – |
| German | 2 | 6 | – | – | – | – |
| Computer Programming | – | – | 1 | 4 | 1 | 4 |
| Electives: | | | | | | |
|    Free | 3 | 10 | 3 | 9 | 3 | 9 |
|    Technical | – | – | 4 | 12 | 9 | 27 |
|    Humanities or Soc. Science | – | – | 6 | 18 | 6 | 18 |
| Totals | 59 | 176 | 42 | 137 | 52 | 167 |

*In 1937–38 some courses ran for more than one term. For purposes of comparison with 1987–88, each two-term course has been counted as two courses.

disappeared, as have the lesser ones in economics, German, and mineralogy. Course requirements in chemical engineering for the baccalaureate degree are double those in the original five-year program; if the present M.Eng. requirements are added, they are three times what they were. The physics requirement is slightly reduced; for mathematics it is somewhat increased; for English it is the same. Computer programming adds four credits to today's requirements. Electives totaled ten credits in 1938; today they total thirty-nine credits for four years and fifty-four for five years. The minimum number of credits for a baccalaureate degree is now 137, compared with 176 fifty years ago. For the present five-year B.S.-M.Eng. program the number is 167.

Simplifying only a little, one can say that much of the chemistry requirement has been replaced by chemical engineering, while electives, many of them technical, have replaced courses in mechanical and electrical engineering.

Some things come full circle. Students graduating in Chemical Engineering before 1943 received two degrees for about 176 credit hours of work; now they can earn the same with 167, only a little less. In 1938 the chemical engineering laboratory course was called just that: Chemical Engineering Laboratory. In 1947 it became Unit Operations Laboratory and so remained until 1986, when it was renamed Chemical Engineering Laboratory. Regardless of its title, it has always been the most demanding and feared course in the undergraduate curriculum.

Research in chemical engineering has changed almost unbelievably. In the 1930's Professors Rhodes and Mason published a few papers on subjects such as "The Crystallization of Paraffin Wax." In the late 1930's Dr. Winding began his studies of heat transfer and of polymeric materials, then a brand-new research area. But research at that time, largely empirical by nature, was considered by Rhodes as secondary to teaching—to be done when time and energy permitted. It is doubtful that $100 a year was spent on chemical engineering research. Even so, this was about as much as was being spent by the rest of the College of Engineering. Research in engineering began under Dean Hollister in the 1940's, but it wasn't until the 1960's and 1970's that it grew explosively, transforming the College.[2] Today chemical engineering research, as noted earlier, is a dominant activity of the School. It has become much more fundamental and less empirical, probing the behavior of atoms and molecules and living cells by sophisticated techniques and computer simulations undreamed of in 1938. Most of the studies are interdisciplinary, on esoteric topics that bear little resemblance to "traditional" sub-

*It is doubtful that $100 a year was spent on chemical engineering research.*

---

[2]See "The Rise of Research" by Julian C. Smith in *Engineering: Cornell Quarterly*, Summer 1985, pp. 26–36.

*Above: The Unit Operations Laboratory around 1960, showing the glass column that exploded.*

*Right above: Another view of the high bay. This is the window a deer once jumped through.*

*Right: Alumni on a tour of Olin during reunion in 1985 heard about plans for better use of the Unit Operations Labaoratory space. Here Professor Robert L. Von Berg is demonstrating the Hewlett-Packard Data Acquisition unit for the heat-transfer experiment.*

jects. The School's early faculty and alumni, if they could see today's research efforts, might well exclaim, "*This* is chemical engineering??"

In 1938 the School was housed in Baker Laboratory. Its total facilities consisted of two small faculty offices, one with an anteroom for the secretary, and a crowded basement laboratory containing worn-out equipment. "It was not only junk," wrote Professor Rhodes, "it wasn't even good junk." Even these cramped quarters were allotted only grudgingly to the new School by the Department of Chemistry. Then, a few years later, came the move to the new building with its spacious offices and lecture rooms and a library and a machine shop and lots of storage space. It's easy to see why Dusty Rhodes designed Olin Hall so expansively—he had dreamed for years of having enough room. His building served the School well for over thirty years, until the growth of research created demands for different spaces, differently arranged. The renovations of the 1970's were a beginning in this direction; now a complete rebuilding of the east wing is under way. The large vacant space in the old Unit Operations laboratory will finally be put to use.

Ever since the move to Olin Hall in 1943, the School has had its own machine shop and wood-working shop. For most of the time it has had two full-time mechanics (later called "mechanicians" and now known as "laboratory equipment technicians"), at least one of whom has been an expert machinist. The first was H. C. Scott, mentioned earlier, who was joined in 1948 by Verne McKinney. When "Scotty" died in 1952, his place was taken by Russell Bush; Russ died in 1963, two years after Verne McKinney retired. Joseph Solomon and Norman Vantine were both employed in 1963. In 1980 Joe retired and Brian Ford was hired in his place; Norm and Brian are currently on the staff. Several of the test units in the Unit Operations Laboratory, as well as many pieces of research equipment, were built by one or the other of these six people.

The Unit Operations laboratory has its own history of unusual events. Fortunately, it has seen no serious accidents; the worst was when a graduate assistant's arm was burned, not badly, by flaming solvent ignited by static electricity. But there have been some very near misses. One time a student was reading a glass thermometer set in the high-pressure steam line; the packing gland let go and the thermometer sailed across the room and shattered against the wall twenty feet away. It missed the student's ear by millimeters. On another occasion two fifth-year students were examining the behavior of a 6-inch glass distillation column near the southeast corner of the laboratory, boiling up a mixture of benzene and toluene. The water-cooled condenser on the column was vented to the room, but inexplicably the vent line had a valve in it. In accordance with Murphy's Law, the vent valve was closed when the steam to the kettle was turned on. There were a few quiet moments while pressure built up; then the column exploded, scattering packing and fragments of glass all over the laboratory and the first-floor hallway as far as Ellie Moulton's office, over 150 feet away. Luckily, the force of the explosion went upward, above the heads of the students. Nobody was hurt.

Other incidents involved animals. In 1960 or thereabouts a loud crash brought everyone running. A female deer, lost and frightened, had jumped through the large glass window near the west end of the laboratory and fallen twenty feet to the concrete floor. Amazingly, the animal was able to stand up and try to run. Russ Bush, the mechanic, wrestled it to the floor. After much discussion, he and the game warden, following legal regulations, took the deer—obviously badly hurt—and released it in a forest. According to Russ, it walked about a hundred yards and fell over and died.

Some time during the 1970's, a squirrel built a nest in the overhead crane and brought up her family there. The crane was not used for several weeks to keep from disturbing the babies. During her stay in Olin Hall, the squirrel lived very well—she had learned how to get inside the candy machine in Room 114 (which now houses the VAX computer) and could gorge herself on chocolate bars.

One final animal story: Once during the Unit Operations experiment in distillation, the column didn't seem to be running quite right. Someone looked through the sight glass and there, clearly visible, was a small yellow rubber duck swimming happily in the pool of liquid on an upper plate. No one seemed to know how it got there. It took a lot of work to get it out. It must also have taken

*Alumni breakfast, 1982*

a lot of work to get it in, unless—as the students said—it really had hatched inside the column.

* * *

Perhaps the biggest change in fifty years is in the relationship of the School to the College of Engineering and the rest of the University. In 1938 the Schools of Civil, Mechanical, and Electrical Engineering were largely independent entities—they had been forcibly joined to form the College only seventeen years earlier. Chemical Engineering, the newcomer, was, if anything, even more independent. It admitted its own students; it set its own curricular requirements and academic standards, which were higher than those in the other schools; it had a five-year baccalaureate program, while the other schools had four-year programs. Soon after its founding, it had a new building, distant from Baker Laboratory and from the other engineering buildings, which were Lincoln, Sibley, and Franklin (now Tjaden) Halls. The School also had extra money: at Dusty Rhodes' urging, alumni set up special funds for student loans, extracurricular activities, and other purposes—funds that Dusty kept well hidden from the College and University administration so he could use them as he saw fit. Chemical Engineering had few ties with its progenitor, Chemistry, and even fewer with other departments and colleges at Cornell. It was pretty much a well-guarded island, an independent state.

Now, of course, it's hard for the School to make any move without infringing on some College or University regulation, some restriction imposed by accrediting agencies, or some governmental dictum. Curricula are tightly constrained. Accounts are monitored by unseen computer operators; no longer can funds be hidden for discretionary use by the director. Committees and committee reports have proliferated. Chemical Engineering is very much part of the College of Engineering, differing in some ways in its undergraduate curriculum and research needs, but otherwise similar to the other schools and departments. It also has close ties through collaborative research to Chemistry and other departments at Cornell and to many other universities around the world.

What happens now? What will chemical engineering be like in the twenty-first century? Will the School of Chemical Engineering at Cornell continue to flourish, or will it go the way of Metallurgical Engineering? Any answers to these questions are little more than speculation, for the only real answer is "No one knows." Still, it's fun to hazard some guesses as to the shape of things in 2038, when some future historian writes about the centennial of the School.

The public image of the chemical engineering profession has had many changes, from puzzled respect fifty years ago to mild admiration in the 1970's to today's negative picture. Even the current antipathy is not universal, for many people recognize that

*". . . many people recognize that problems like air and water pollution, acid rain, and toxic-waste disposal can best be solved by chemical engineers."*

*The 1988 faculty photograph. First row, left to right: Robert P. Merrill, Helen Downs Haller, Peter A. Clark, Keith E. Gubbins, Paulette Clancy. Second row: Clarence Shoch (visiting professor from Du Pont), Ferdinand Rodriguez, William B. Streett, Robert L. Von Berg, Robert K. Finn, Herbert F. Wiegandt. Third row: Paul H. Steen, John Rowlinson (visiting professor from Oxford University), John A. Zollweg, Daniel A. Hammer, Peter Harriott. Top row: Donald L. Koch, Michael L. Shuler, A. Brad Anton, George F. Scheele, Athanassios Panagiotopoulos, Claude Cohen.*

problems like air and water pollution, acid rain, and toxic-waste disposal can best be solved by chemical engineers. When petroleum once again runs short, chemical engineers will be in great demand to develop synthetic fuels. The 1988 Amundson report to the National Research Council lists many promising research areas for chemical engineers, including biotechnology, electronic devices, microstructured materials, in-situ processing of resources, liquid fuels, management of hazardous materials, process control, and surface and interfacial engineering. It also discusses the continuing opportunities in "classical" chemical engineering—in petroleum refining, chemical-intermediates processing, and the production of key large-volume chemicals. More public awareness of the work of chemical engineers should lead to better understanding and a much more favorable image.

The School of Chemical Engineering at Cornell has survived some very trying times. It shows no sign of disappearing; on the contrary, it is vigorous and growing. The undergraduate and professional master's programs change but continue strong. The balance between undergraduate and graduate studies is, of course, very different from what it was—the School is now a research-oriented unit in a research-oriented College. Does this mean that in fifty years there will be no undergraduate program? I doubt it. I think there will be a strong group of B.S. graduates in the class of 2038. But what they will have studied and what kind of work they will do are hard to imagine. They may be going out from an Olin Hall that is once again run-down and unsuited to the needs of the School—after all, the east wing of the building was rebuilt long ago in 1988, and even the north wing was last renovated in 2003.

The only certainty is that things will change, sometimes for the better, sometimes not. The story of the School will go on. To bring this one to a close, it seems appropriate to quote (with slight modifications) the words Dusty Rhodes wrote thirty years ago at the end of his 1957 history:

*The School now belongs to the younger faculty; it is now their responsibility to insure that it is used with the utmost effectiveness to train young men and women for the greatest possible proficiency in their chosen profession and the greatest possible service to society.*

> *I think there will be a strong group of B.S. graduates in the class of 2038. But what they will have studied and what kind of work they will do are hard to imagine.*

# Appendices

Appendix A-1

## Degrees Held by Faculty Members

| Name | Bachelor's | Master's | Doctorate |
|---|---|---|---|
| Fred H. Rhodes | Wabash College '10 | | Cornell '14 |
| Charles C. Winding | Minnesota '31 | | Minnesota '35 |
| Clyde W. Mason | Oregon '19 | | Cornell '24 |
| Oscar J. Swenson | Minnesota '31 | | Minnesota '35 |
| Julian C. Smith | Cornell '41 | Cornell '42 | |
| Robert L. Von Berg | West Virginia '40 | West Virginia '41 | MIT '44 |
| Herbert F. Wiegandt | Purdue '38 | Purdue '39 | Purdue '41 |
| Jay E. Hedrick | Illinois College '31 | Iowa '32 | Iowa '34 |
| Raymond G. Thorpe | RPI '42 | Cornell '47 | |
| Peter Harriott | Cornell '49 | | MIT '52 |
| Robert K. Finn | Cornell '41 | Cornell '42 | Minnesota '49 |
| Robert York | Tennessee '33 | Tennessee '34 | MIT '38 |
| Ferdinand Rodriguez | Case Institute '50 | Case Institute '54 | Cornell '58 |
| George F. Scheele | Princeton '57 | Illinois '59 | Illinois '62 |
| George G. Cocks | Iowa State '41 | | Cornell '49 |
| Jean Paul Leinroth Jr. | Cornell '41 | MIT '48 | MIT '63 |
| Lemuel Wingard | Cornell '53 | | Cornell '65 |
| Victor Edwards | Rice '62 | | Berkeley '67 |
| David M. Watt Jr. | Princeton '64 | | Berkeley '69 |
| Kenneth B. Bischoff | IIT '57 | | IIT '61 |
| John L. Anderson | Delaware '67 | Illinois '69 | Illinois '71 |
| James F. Stevenson | RPI '65 | Wisconsin '67 | Wisconsin '70 |
| Michael L. Shuler | Notre Dame '69 | | Minnesota '73 |
| Keith E. Gubbins | London '58 | | London '62 |
| Robert P. Merrill | Cornell '60 | | MIT '64 |
| Claude Cohen | American U., Cairo '66 | | Princeton '72 |
| Joseph F. Cocchetto | Cornell '73 | MIT '75 | MIT '80 |
| William L. Olbricht | Stanford '73 | | CalTech '80 |
| William B. Streett | U.S. Military Acad. '55 | Michigan '60 | Michigan '63 |
| Paul F. Steen | Brown '75 | | Johns Hopkins '81 |
| Paulette Clancy | London '74 | | Oxford '77 |
| Douglas S. Clark | Vermont '79 | | CalTech '83 |
| A. Brad Anton | VPI '79 | CalTech '82 | CalTech '85 |
| Athanassios Z. Panagiotopoulos | Nat'l Tech. U., Athens '82 | | MIT '86 |
| Donald L. Koch | Case Western Reserve '81 | | MIT '86 |
| Peter A. Clark | Clarkson '78 | Carnegie-Mellon '80 | Carnegie-Mellon '83 |
| John A. Zollweg | Oberlin College '64 | | Cornell '69 |
| Daniel A. Hammer | Princeton '82 | Pennsylvania '85 | Pennsylvania '87 |

Appendix A-2

# Faculty Appointments

| Name | Instructor | Assistant | Associate | Professor | Chair (and year appointed) | Director | Retired | Resigned | Died |
|---|---|---|---|---|---|---|---|---|---|
| Fred H. Rhodes | 1915 | - | - | 1920 | Herbert F. Johnson '41 | 1937-57 | 1957 | 1917 | 1976 |
| Charles C. Winding | 1935 | 1938 | 1941 | 1944 | Herbert F. Johnson '57 | 1957-70[a] | 1975 | - | 1986 |
| Clyde W. Mason | 1924 | 1928 | - | 1933 | Emile M. Chamot '58 | - | 1966 | - | 1983 |
| Oscar J. Swenson | - | 1938 | 1941 | - | - | - | - | 1946 | - |
| Julian C. Smith | - | 1946 | 1949 | 1953 | - | 1975-83[b] | 1986 | - | - |
| Robert L. Von Berg | - | 1946 | 1949 | 1958 | - | - | - | - | - |
| Herbert F. Wiegandt | - | 1947 | 1949 | 1952 | - | - | 1987 | - | - |
| Frank Maslan | - | 1947 | - | - | - | - | - | 1948 | - |
| Jay E. Hedrick | - | - | - | 1949[g] | - | - | 1975 | - | 1981 |
| Raymond G. Thorpe | - | 1951 | 1954 | 1984 | - | - | - | - | - |
| Peter Harriott | - | 1953 | 1954 | 1966 | Fred H. Rhodes '75 | - | - | - | - |
| Robert K. Finn | - | - | 1955 | 1961 | - | - | - | - | - |
| Robert York | - | - | - | 1956 | Socony-Mobil '56 | - | - | - | 1978 |
| Ferdinand Rodriguez | - | 1958 | 1962 | 1971 | - | - | - | - | - |
| George F. Scheele | - | 1961 | 1969 | 1987 | - | [c] | - | - | - |
| George G. Cocks | - | - | 1964[e] | - | - | - | - | 1981 | - |
| Jean Paul Leinroth Jr. | - | - | 1964[d] | - | - | - | - | 1971 | - |
| Lemuel Wingard | - | 1965[d] | - | - | - | - | - | 1966 | - |
| Victor Edwards | - | 1967 | - | - | - | - | - | 1971 | - |
| David M. Watt Jr. | - | 1969 | - | - | - | - | - | 1971 | - |
| Kenneth B. Bischoff | - | - | - | 1970 | Walter R. Read '70 | 1970-75 | - | 1976 | - |
| John L. Anderson | - | 1971 | - | - | - | - | - | 1976 | - |
| James F. Stevenson | - | 1971 | 1977 | - | - | - | - | 1977 | - |
| Michael L. Shuler | - | 1974 | 1979 | 1984 | - | - | - | - | - |
| Keith E. Gubbins | - | - | - | 1976 | Thomas R. Briggs '76 | 1983- | - | - | - |
| Robert P. Merrill | - | - | - | 1976 | Herbert F. Johnson '76 | - | - | - | - |
| Claude Cohen | - | 1977 | 1981 | 1987 | - | - | - | - | - |
| Joseph F. Cocchetto | - | 1980 | - | - | - | - | - | 1985 | - |
| William L. Olbricht | - | 1980 | 1986 | - | - | - | - | - | - |
| William B. Streett | - | - | - | 1981[h] | - | - | - | - | - |
| Paul F. Steen | - | 1982 | - | - | - | - | - | - | - |
| Paulette Clancy | - | 1984[f] | - | - | - | - | - | - | - |
| Douglas S. Clark | - | 1984 | - | - | - | - | - | 1986 | - |
| A. Brad Anton | - | 1987 | - | - | - | - | - | - | - |
| A. Z. Panagiotopoulos | - | 1987 | - | - | - | - | - | - | - |
| Donald L. Koch | - | 1987 | - | - | - | - | - | - | - |
| Peter A. Clark | - | 1987 | - | - | - | - | - | - | - |
| John A. Zollweg | - | - | 1987[d] | - | - | - | - | - | - |
| Daniel A. Hammer | - | 1988 | - | - | - | - | - | - | - |

[a] Assistant Director 1947-57  
[b] Associate Director 1973-75  
[c] Assistant Director 1982; Associate Director 1983-  
[d] Without tenure  
[e] Without tenure; granted tenure 1968  
[f] Without tenure; tenure track 1987-  
[g] Assistant Dean 1952-55  
[h] Senior Research Associate 1978; Dean of Engineering 1985-

Appendix A-3

# **Faculty: Periods of Service**

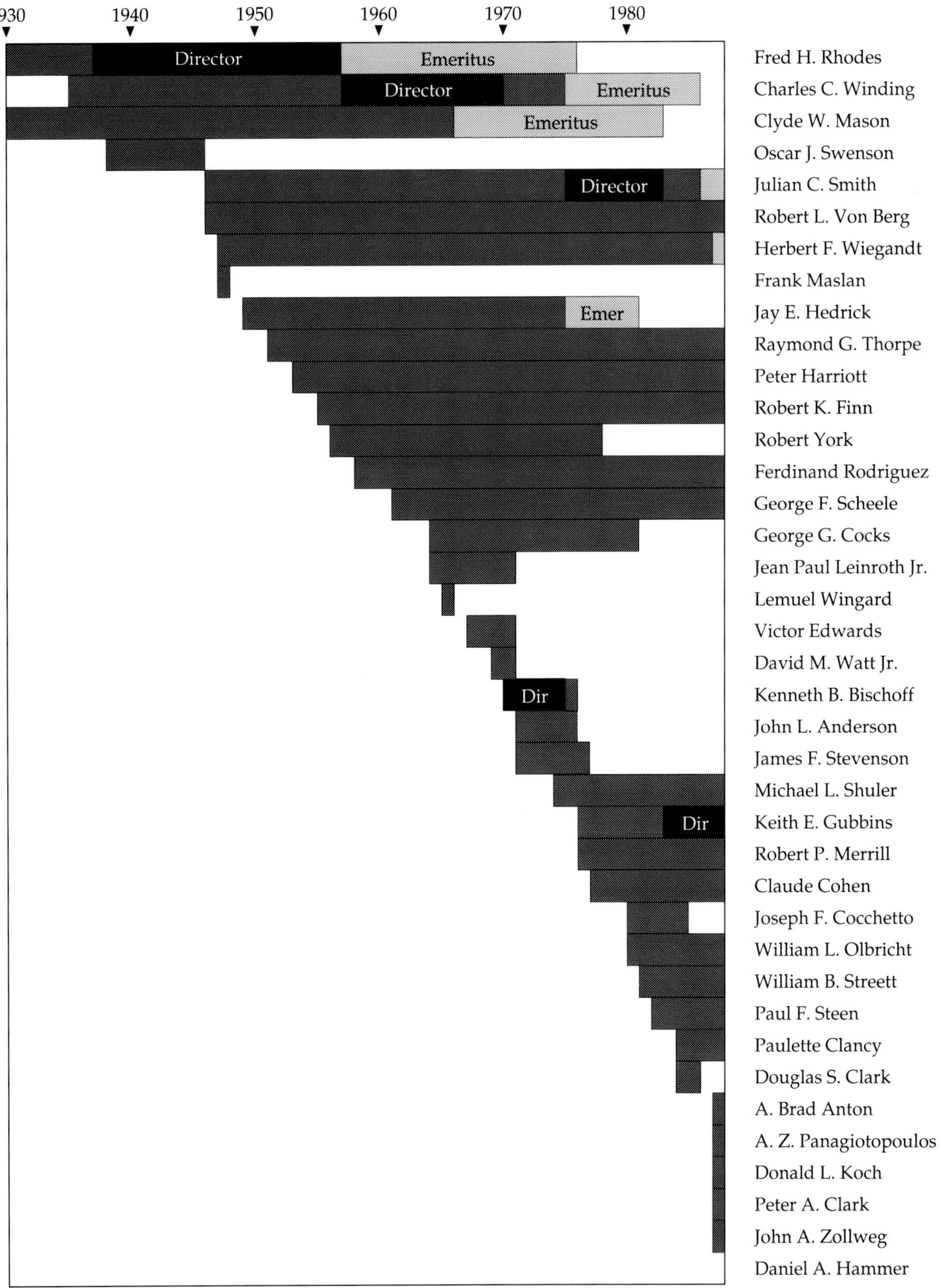

Appendix B

# **Faculty Memorial Statements**
from the official Cornell University records


### FRED HOFFMAN RHODES
June 30, 1889 – November 30, 1976

Fred Hoffman "Dusty" Rhodes, affectionately known as the father of chemical engineering at Cornell, died November 30, 1976 in De Land, Florida. He was born June 30, 1889 in Rochester, Indiana, where he completed his elementary education, graduating from high school in 1906. He then entered Wabash College, where he majored in chemistry, also acting as an English instructor in his senior year. Apparently this was the start of his interest in perfection in writing that later plagued many Cornell chemical engineering students but proved to be of great help to them in their professional careers.

The Cornell Department of Chemistry needed an assistant in qualitative analysis in February 1910. Dusty accepted the job, although he had never had qualitative analysis at Wabash. He later became a personal research assistant to Louis Dennis, the head of the Department of Chemistry for many years.

After receiving his Ph.D. degree from Cornell in 1914, he went to the University of Montana for a year to teach chemistry and metallurgy. In 1915 he returned to Cornell as an instructor in qualitative analysis, the course he had never taken. After two years he decided he needed industrial experience, and from 1917 to 1920 he worked for the Barrett Company, starting out as a research chemist and ending up as director of research. In this period he contributed to some of the developments that became the foundation of chemical engineering. He returned to Ithaca and Cornell in 1920 as a professor of industrial chemistry. At this time, Professor Dennis was buying equipment in Europe for Baker Laboratory, which was then under construction, leaving Dusty with instructions to buy some equipment that might be suitable for industrial chemistry. When Dennis returned, he found that some of the equipment that had been installed in the basement of Baker Laboratory was curiously similar to that found in chemical engineering laboratories.

For the next ten years there is no documented evidence that chemical engineering as such at Cornell was anything but a figment of Dusty's imagination. During this period, however, Cornell graduated many bachelors of chemistry, who later turned out to be some of the outstanding leaders of the chemical industry. It was also during this period that Dusty published a number of articles covering such things as soaps, lubricating oils and greases, phenol and paints. Interspersed among them were research papers on unit operations. In the worst part of the depression he finally convinced the faculty that there was such a thing as a chemical engineer, at least to the extent that they agreed the degree of chemical engineer would be granted to any bachelor of chemistry who completed a fifth year under Dusty's direction. In 1933 the first class of three chemical engineers was graduated.

In 1938 the School of Chemical Engineering was created as a separate school in the College of Engineering with an integrated five-year course leading towards the degree of bachelor of chemical engineering. At this time the school's faculty consisted of Rhodes and an assistant professor, with an occasional instructor when one could be found who would work hard enough to meet Dusty's standards. The official faculty, however, included two other engineers and two chemists carefully chosen so that faculty policies did not conflict with those of Professor Rhodes. This proved to be very satisfactory, and before long it became unnecessary to hold faculty meetings.

Since Dusty had achieved his objective and had created a separate School of Chemical Engineering, the next step involved obtaining a suitable building to house the school. Fortunately, S. C. Hollister, the dean of the College of Engineering at that time, proved an able and willing coworker in obtaining a chemical engineering building, which was to be the first unit of the new engineering quadrangle. In 1940, Franklin W. Olin donated the funds for Olin Hall, and construction was started early in 1941. The new Olin Hall of Chemical Engineering was first used in May of 1942, and Dusty was named the first Herbert Fisk Johnson Professor of Industrial Chemistry.

During the World War II period, in addition to a heavy twelve-month teaching load, Rhodes served in the Office of Production Research and Development, under the War Production Board. He was also developing staff, facilities, and a curriculum for metallurgical engineering, a new discipline for Cornell, and a bachelor's degree program was started in 1947. The school then became the School of Chemical and Metallurgical Engineering until 1963, when metallurgical engineering was combined with materials science. In addition, Dusty was elected a director of the former German firm, the General Aniline and Film Corporation, a post he held for nearly ten years. The influx of veterans after the war caused a critical housing shortage for Cornell, threatening to lower the number of chemical engineering students, so Dusty provided rooms for twenty in Olin Hall under strict rules to govern their behavior.

Dusty officially retired July 1, 1957, after a year's terminal sabbatic leave (the only one he ever took) to "go fishing." He then proceeded to write a history of the chemistry department and the chemical engineering school and was elected an alumni trustee for a five-year term. Shortly afterwards, the Cornell Alumni Association voted that all candidates for this position must have been a Cornell undergraduate, which Dusty was not.

Dusty, above all, insisted his chemical engineering students be given the best possible chance to achieve the competence needed to further their careers. He required excellence in teaching; he helped provide facilities and financial support for the school and for chemical engineering students; he strongly resisted interference by outsiders and had the personality to succeed in these endeavors. Dale R. Corson, president of Cornell University and former dean of engineering, made the following comments:

"Dusty Rhodes was himself no ordinary person, and he wanted extraordinary individuals as students. He wanted to teach and train superior engineers. With a humanity covered with a veneer of gruffness and mild chicanery, he built the curriculum and the program, forced his students to superior work, and then assured them of positions of status in the profession. He fought for his students, he supported them, and he defended them against incursions from alien beings. He continued to be concerned about them when they left Olin Hall. His continued interest in the fate and fortunes of chemical and metallurgical alumni is well documented."

About a year before Rhodes retired, a small group of his former students and Professor Winding of the School of Chemical Engineering, formed a committee to attempt to raise money from the then approximately seven hundred chemical engineering alumni to endow a chemical engineering professorship in Rhodes's name. It was an ambitious undertaking for such a small group, but by 1970 well over half a million dollars was accumulated, almost all of it from Dusty's former students. In 1971 the Fred Hoffman Rhodes Professorship in Chemical Engineering was established. Dusty was very pleased when one of his students, Professor Peter Harriott, was made the first holder of this professorship. The professorship is a fitting tribute to an extraordinary person, an example of the affection and high regard extended by his students.

Peter Harriott<br>
Franklin A. Long<br>
Julian C. Smith<br>
Charles C. Winding

CHARLES CALVERT WINDING
August 12, 1908 - March 17,1986

Charles C. (Chuck) Winding died on March 17, 1986, after a lifetime of dedicated teaching and service to Cornell. With Fred H. ("Dusty") Rhodes he was a founder of the School of Chemical Engineering, and over a fifty-year period he saw it prosper, grow, and change. For thirteen of those years he was the director of the school, leading it successfully through some very difficult times.

Chuck was born in Minneapolis and received his B.S. and Ph.D. degrees from the University of Minnesota. In 1935 Dusty Rhodes asked him to join, with the rank of instructor, in the development of chemical engineering at Cornell. Chuck accepted and moved to Ithaca, where he spent the rest of his life. In 1936 he married Katherine (Kay) Cudworth, who survives him; he had met her at the University of Minnesota. In 1938 he was appointed assistant professor of chemical engineering. He became associate professor in 1941, a professor in 1948, and the Herbert Fisk Johnson Professor of Industrial Chemistry in 1957. That year he also became director of the School of Chemical Engineering, a position he held until 1970.

Professionally Chuck's first love was teaching. Always friendly and helpful to students, he nonetheless demanded hard work, correct answers, and an appreciation of the breadth and diversity of chemical engineering practice. His weekly quizzes, graded 0 or 10, created near-panic among fifth-year students in the 1940s. "An engineering answer has to be right," he said. "A bridge that's *almost* long enough isn't worth anything." He taught courses in chemical process design and organized several more in the developing field of polymer technology. Shortly before he retired, he restructured and modernized the undergraduate process design course and continued teaching it, on a part-time basis, for three years after he became professor emeritus.

He was a strong proponent of the five-year bachelor's degree program in engineering, believing that four years was not sufficient to give students adequate preparation for professional practice. But when the college abandoned the five-year program in 1965, he adapted well to the situation and saw to it that the strength and vitality of the chemical engineering program was not lost in the new curriculum. He had previously developed a professional engineering program in chemical engineering at the master's level, which in large measure replaced the fifth year of the old program.

Chuck carried a heavy teaching load, especially early in his career, when he and Dusty Rhodes taught most of the courses in chemical engineering. In 1940-42, while Rhodes was occupied with the construction of Olin Hall, Chuck taught nearly all the courses by himself while still contributing to the plans for the new building. During the war years his load increased still further, for under the accelerated schedule there were three terms each year and all courses were given each term. There were no vacations. When asked about that he merely said dryly, "I was glad when it was over."

As a researcher, he was meticulous and thorough, skilled in asking the right questions and in devising experimental techniques to answer them. He never spared himself time or effort in his research work and was equally demanding of his graduate students. His fundamental studies of heat transfer featured truly innovative techniques, and in polymer studies, starting with his Ph.D. thesis on cellulose acetate, he was one of the pioneers. During World War II his group studied non-Newtonian flow and degradation of rubber latexes and solutions. The course he offered in synthetic plastics, beginning in 1943, was one of the first courses on polymers given in a chemical engineering school. For years he wrote the annual review of polymerization for Industrial and Engineering Chemistry. He also wrote two books on polymer technology — one with Leonard Hasche of Tennessee Eastman (1947), the other with Gordon Hiatt of Eastman Kodak (1961). His biographical profile was included recently in "Polymer Pioneers" in Polymer News.

In 1957 he became director of the school, following Dusty Rhodes's colorful career. He was more conservative than his predecessor but maintained Rhodes's policies of concern for the professional development of his students. He did not favor the growing emphasis on research in the engineering college, despite his own prowess in research, since he believed it would inevitably reduce the attention paid to undergraduate education. Chuck led the school through the turbulent 1960s. He battled successfully against the forces that would have significantly reduced the amount of chemistry in the chemical engineering curriculum, modified and shortened the bachelor's program from five to four years, and held everything together during the student disruptions of 1968-1970. During that difficult period he established several new and much-needed options in the chemical engineering curriculum.

For almost twenty years he wrote the *Olin Hall*

## CLYDE WALTER MASON
### June 17, 1898 - December 8, 1983

*News*, a newsletter to the alumni. He knew virtually all of the graduates personally and kept in close touch with many of them through the newsletter and at annual meetings of various technical societies. He was a fellow of the American Institute of Chemical Engineers and a distinguished member of the Society of Plastics Engineers. In 1983 he was named Educator of the Year by that society, and a few days before he died he received a certificate celebrating fifty years of membership in the Americal Chemical Society.

During Word War II Chuck was a consultant to the Office of Rubber Reserve. Later he consulted with Rome Cable and a number of other companies. For many years he was a director of the Cowles Company in Skaneateles.

He was an avid sailor, competing with great success in meets and regattas on Seneca and Cayuga Lakes. His sailboat, a Thistle, was appropriately named Poly-Mer. He was a past commodore of the Ithaca Yacht Club, and for many years secretary-treasurer of the Central New York Yacht Racing Association.

Chuck was a dedicated, sincere, conscientious, and caring gentleman with a host of friends. In 1973 a dinner organized by chemical engineering alumni was held in Philadelphia to recognize his many contributions to the School of Chemical Engineering and to the education and welfare of his students. Announced at that dinner was the establishment of the Charles C. Winding Scholarship Fund, made possible by contributions from alumni and friends. Over the years additional gifts have swelled this fund considerably. It will continue to keep his memory alive as it supports graduate students in the program he liked best of all — the professional master's degree program in chemical engineering.

*Ferdinand Rodriguez*
*Raymond G. Thorpe*
*Robert L. Von Berg*
*Julian C. Smith*

Clyde Mason was born in Watertown, South Dakota, in 1898. He went to college in Eugene, Oregon, and received an A.B. degree in chemistry from the University of Oregon in 1919. He stayed on there for a year of graduate study but in 1920 seized an opportunity to combine his interest in chemistry with his love of microscopes and came to Cornell as a Ph.D. candidate under Professor Emile M. Chamot, with a major in chemical microscopy. Microscopy was not his only love, however, for during his first year at Cornell he met and married Elizabeth M. Peterson. For the next sixty-three years Ithaca and Cornell were their home.

In 1924 he received his doctorate and was appointed instructor in chemistry at an annual salary of $1,200. He was made an assistant professor in 1927 and a professor in 1933. With his colleague and former teacher E. M. Chamot he published the first edition of the classic Handbook of Chemical Microscopy in 1930-31. When Olin Hall was completed in 1942, providing new facilities for teaching and research in microscopy, Clyde moved out of Baker Laboratory and became professor of chemical microscopy and metallography in the School of Chemical Engineering. In 1958 he was named the Emile M. Chamot Professor of Chemical Microscopy. He retired from teaching in 1966 but continued working at Olin Hall until a few days before he died. His revised fourth edition of Volume 1 of the Handbook was published in 1983, when he was eighty-four.

This brief catalog of events, however, does little to convey a sense of his remarkable character. His sometimes prickly independence and self-sufficiency reflected his western origin. He was devoted to learning and to truth, constantly studying, criticizing and writing with uncompromisingly high standards. His vast knowledge, always up-to-date, made him a true authority in his field. He knew a great many other things, too - most unexpected things. "Ask Clyde," one would say when a seemingly impossible question arose. "He'll know." And he nearly always did.

His great love, however, was teaching. "I knew from the beginning that I was not a great lecturer," he said, "so I concentrated on becoming a good teacher. I particularly liked the beginner or struggler." This meant that he liked just about everybody, for all the students in his classes did a lot of struggling. He expected budding engineers to share his passion for accuracy and independent observation. "Don't tell me what the book says," he would tell the weary-eyed student bent over a microscope. "Tell me what *you* see." He abhorred the parroting of undigested information. "Don't take notes on my lecture," he would say. "Listen to it - think about it." Shocked and dismayed, most students would take notes anyway, for they had never before been expected to think about lecture material during class. His examinations were equally searching, and he was a tough grader. Nonetheless, he was held in high esteem by his students and was sensitive to their opinions. He especially appreciated the Christmas gift from one of the classes of some notepads inscribed, "From the little world of C. W. Mason."

Clyde's world was anything but little. He wrote some forty technical articles on microscopy applied to a wide range of subjects. A series of lectures he presented to the American Society for Metals was published by that society in 1947 as a book entitled Introductory Physical Metallurgy. He was a consultant to several industrial firms, the Army Chemical Corps, and the Office of Scientific Research and Development. He was the founder and first chairman of the Division of Analytical and Microchemistry of the Americal Chemical Society; he was a fellow of the New York Microscopical Society and an active member of several other technical societies. He had a keen interest in libraries, served on library committees, and taught a course in library use. For many years he served on the University Committee on Music.

During the 1960s he drove a Model A Ford that he had lovingly restored to "historic vehicle" quality; he estimated that he spent three thousand hours on the project. He and his wife loved to dance, and he organized frequent dances for the Alpha Chi Sigma fraternity, of which he was faculty adviser. He was an excellent figure skater and an early member of the Cornell Figure Skating Club. Here he revealed his extreme fondness for children, a characteristic known only to his closest associates. He somehow arranged to be at Lynah Rink whenever he suspected that a chemical engineering faculty child would be at a public skating session. Parents were summarily dismissed; he made sure that blades were properly sharpened, that boots were properly laced, and that the child was not rigidly supported while learning to "use the edges." At age seventy he still had the patience to teach a four-year-old to skate.

Formal honors came to him rather late in life. In 1969 he was cited by the New York Microscopical Society for his contributions, and in 1981 an alumnus of the Class of 1956 established an engineering scholarship in his name. Currently three Master of Engineering degree students are designated as Mason Scholars.

For many years Clyde lived in Cayuga Heights with his wife and two children, George and Phoebe, all of whom survived him. Independent as always, Clyde refused outside assistance, even when, near the end of his days, he was caring for his increasingly incapacitated wife. "We'll manage," he would say. But the day finally came when even he could manage no longer.

Clyde was a gifted, exacting, complex individual. He was a distinguished gentleman, reserved yet friendly; outwardly brusque yet sensitive and unfailingly generous; demanding of the students yet deeply concerned with their welfare; a recognized authority of enormous learning, yet one who often apologized for his lack of knowledge. There was no pretense in him. Above all he was a devoted teacher who well understood the learning process and the limitations of our educational system, who sought to awaken in his students a sense of independent thought and critical judgment. Cornell has lost a dedicated servant. It is comforting that he lived to see the establishment of the Mason Engineering Scholarship that will perpetuate his memory.

Robert L. Von Berg<br>Charles C. Winding<br>Julian C. Smith

JAY ELDRED HEDRICK
July 17,1909 - June 10, 1981

Jay Hedrick, professor emeritus of chemical engineering, died on June 10, 1981. He was born in 1909 in Meredosia, Illinois; he received his Bachelor of Arts degree from Illinois College, and his Master of Science and Doctor of Philosophy degrees in chemical engineering from the State University of Iowa at Ames. Jobs were scarce in 1934, even for a Ph. D. chemical engineer, but Jay found employment with the Iowa Coal Laboratory, with the Iowa Public Health Department, and then, for five years, as instructor in chemical engineering at Kansas State College, where he directed research projects on petroleum and coal. Many of the present-day proposals for coal utilization were studied and evaluated by Jay in the late 1930s.

A few months before Pearl Harbor, Jay left Kansas to join Shell Oil in San Francisco as technical supervisor (and later as senior engineer and senior technologist). For eight months in 1944-45 he was on leave from Shell to work at the War Production Board in Washington. At the end of the war he was very happy to return to San Francisco; so when Shell moved its offices to New York City in 1949, Jay reluctantly went along but began to look for something new - preferably an academic position in a semirural area.

At this time, Fred H. "Dusty" Rhodes, director of the School of Chemical Engineering, was looking for a senior professor with industrial experience. Once Jay learned of this, things moved quickly. In September 1949 he joined the chemical engineering faculty, bought a house in Cayuga Heights, and became an Ithacan for the rest of his life. Jay claimed that what Dusty really wanted was someone to join (and be fleeced at) his weekly poker sessions: "I fitted right in," said Jay. "I was a lousy poker player."

Jay taught a variety of chemical engineering courses during his twenty-six years at Cornell, mostly in process economics, commercial development, and chemical product marketing. He served on numerous committees of the school, college, and University, among them the Centennial Planning Committee, the University Council, and the University Lectures Committee. For some years he was faculty adviser to the student chapter of the American Institute of Chemical Engineers. From 1953 to 1956 he was assistant dean of the College of Engineering ("I was a mouse," he said, "learning to be a rat."); but when Dean S. C. Hollister retired, Jay elected to leave administration and return full-time to the less rodentlike world of teaching.

For many years Jay was a consultant to Shell Oil and other organizations, chiefly on matters of product development, commercialization, and marketing. He retired in July 1975 and was named professor emeritus, but he continued to come to his office in Olin Hall nearly every day and kept up an active program of consulting and research. He was a member of the American Institute of Chemical Engineers, the American Chemical Society (chairman, Cornell Section, 1952), Alpha Chi Sigma, Phi Lambda Upsilon, Sigma Xi, and Tau Beta Pi; and a fellow of the American Institute of Chemists. He is listed in Who's Who in America.

His first wife, Mary Ellen, died in 1957, leaving him with four children - a boy and three girls. A few years later he married Betty Cook and had another daughter when he was fifty-five. He once threatened to write a book called "My Fifty Years in the PTA." When he died he had eleven grandchildren, all girls.

Phrases that come to mind in describing Jay are warm-hearted, friendly, informal, helpful, knowledgeable, and keenly interested in current affairs. He went out of his way to help young professors get started. He always tore up his lecture notes when a course was over, to insure that next year's would be up to date. He never lost interest, even after retirement, in the condition and potential developments of chemical commodity markets. Toward the end of his life he assembled detailed genealogical information about his family for presentation to his children. He was an excellent conversationalist: he loved to talk, and did so with authority, on a wide range of subjects. He was also noted for his delightful sense of humor.

But most outstanding of Jay's qualities was his courage. Beset by cancer even before he retired, he endured four major operations and recovered remarkably from them all. He was always cheerful and forward-looking, never gloomy; he exercised faithfully and, with his wife Betty, rode his bicycle several miles a day around the streets of Cayuga Heights. His attitude toward his illness was extraordinary.

Jay Hedrick was an effective teacher, a knowledgeable researcher and consultant, a respected colleague, a truly good neighbor, and a beloved and loving husband and father. We will all miss him greatly.

*Blanchard L. Rideout*
*Charles C. Winding*
*Julian C. Smith*

82

ROBERT YORK
December 21, 1912 - January 7, 1978

Robert York brought to Cornell a broad and expert background in chemical engineering, derived from earlier academic and industrial experience. For more than twenty years he offered soft-spoken guidance and sensitive encouragement to students, colleagues, and friends in this campus community. A devoted and demanding teacher, Professor York gave his students a broad perspective that combined the technical and theoretical with economic, social, and political practicality. His interest in them was maintained long after graduation; their esteem for him grew with the years.

Robert York's associates regarded him not only as an outstanding engineer and teacher but also as a warm friend. To lunch with him was a delightful break in the daily routine. Bob was part of a luncheon group that originated at a table in the Willard Straight Hall Memorial Room long before Statler Hall was built. Later this group occupied a Rathskeller round table. Composed largely of senior faculty from several schools and colleges, this gathering represented a great diversity in points of view and training. Its discussions were sharp, the arguments sometimes hilarious and always enjoyable. Here his comments carried weight because they arose from an incisive intellect that was not confused by peripheral matters. His ability to get to the heart of a problem was well recognized, not only with this group, but elsewhere. While he served on the University Senate, discussions of University affairs became a favorite and often entertaining subject. His subtle humor was well chosen, and he was not above leading on some unwitting person to the amusement of others. As the group decreased in number with the retirement, removal, or death of many of the original members, Bob became as much of a chairman as was needed. His eager participation led to many close friendships; his death has left an emptiness that cannot be filled.

Professor York was modest about his achievements, though he held memberships in well-known scholastic and professional societies and was listed in highly regarded directories. Major corporations sought his counsel, as did friends, associates, and students. He was active on many University and professional committees. His interests and capabilities extended far beyond his field of specialization to include economic and financial matters, patents, applied mathematics, and marketing.

*Jay E. Hedrick*
*Shailer S. Philbrick*
*Raymond G. Thorpe*

Appendix C

# B.Ch.E. Curricula

| | 1937–38 | F | S | 1947–48 | F | S | 1957–58 | F | S |
|---|---|---|---|---|---|---|---|---|---|
| *Terms 1,2* | | | | | | | | | |
| Inorganic Chemistry | Chem 110 | 3 | 2 | Chem 111,112 | 3 | 2 | Chem 113, 114 | 4 | 4 |
| Chemistry Laboratory | Chem 115 | 3 | - | Chem 115 | 3 | - | | | |
| Qualitative Analysis | Chem 203 | - | 5 | Chem 212 | - | 5 | | | |
| Anal. Geom. & Calculus | Math 5a,5b | 5 | 5 | Math 161,162 | 3 | 3 | Math 161, 162 | 3 | 3 |
| Experimental Physics | Phys 11,12 | 4 | 4 | Phys 115,116 | 3 | 3 | Phys 115, 116 | 3 | 3 |
| English | Engl 2 | 3 | 3 | Engl 111,112 | 3 | 3 | Engl 111, 112 | 3 | 3 |
| Drawing & Desc. Geom. | | | | ME 3114,3115 | 2 | 2 | Engr 3117,3118 | 2 | 2 |
| Orientation | | | | | | | Engr 5000 | 0 | 0 |
| | | **18** | **19** | | **17** | **18** | | **15** | **15** |
| *Terms 3,4* | | F | S | | F | S | | F | S |
| Organic Chemistry | Chem 305 | 3 | 3 | Chem 307,308 | 3 | 3 | Chem 307, 308 | 3 | 3 |
| Organic Chemistry Lab. | Chem 310 | 3 | 3 | Chem 311,312 | 3 | 3 | Chem 312, 312 | 2 | 2 |
| Quantitative Analysis | Chem 220 | 3 | - | Chem 220 | 3 | - | Chem 224 | 4 | - |
| Quantitative Anal. Lab. | Chem 221 | 3 | - | Chem 221 | 3 | - | | | |
| Gas and Fuel Analysis | Chem 250 | - | 3 | | | | | | |
| Physics | Phys 21,22 | 3 | 3 | Phys 117,118 | 3 | 3 | Phys 117, 118 | 3 | 3 |
| German | German 1b | 3 | 3 | | | | | | |
| Engineering Drawing | M.E. 125 | - | 3 | | | | | | |
| Analytics and Calculus | | | | Math 163 | 3 | - | Math 163 | 3 | - |
| Differential Equations | | | | Math 201 | - | 3 | | | |
| Applied Mathematics | | | | | | | Engr 1156 | - | 3 |
| Mechanics | | | | ME 1125 | - | 3 | Engr 1151 | - | 3 |
| Chem. Eng. Stoichiometry (Intro. Chem. Eng.) | | | | ChE 501 | - | 2 | ChE 5101, 5102 | 2 | 2 |
| Public Speaking | | | | P.S. 101 | - | 3 | P.S. 101 | - | 3 |
| (or Psychology) | | | | Psych 440 | - | (3) | | | |
| | | **18** | **18** | | **18** | **20** | | **17** | **19** |
| *Terms 5,6* | | F | S | | F | S | | F | S |
| Physical Chemistry | Chem 405 | 3 | 3 | Chem 403,404 | 3 | 3 | Chem 403, 404 | 3 | 3 |
| Physical Chemistry Lab. | Chem 410 | 3 | 3 | Chem 411,412 | 3 | 3 | Chem 411, 412 | 2 | 2 |
| Mechanics | Eng 3M21,22,23 | 5 | 5 | ME 1126 | 3 | - | Engr 1152 | 3 | - |
| Materials of Construction | Eng 3X21,22 | 3 | 3 | ChE 1255,1256 | 3 | 3 | ChE 6255, 6256 | 3 | 3 |
| Mineralogy | Geol 311 | 3 | - | | | | | | |
| Strength of Materials | | | | ME 1127 | - | 3 | Engr 1153 | - | 3 |
| Science in Western Civilization | | | | Hist 165,166 | 3 | 3 | | | |
| Chemical Engineering Technology | | | | ChE 5203,5204 | 2 | 2 | ChE 5203, 5204 | 2 | 2 |
| Unit Operations of Chemical Engineering | | | | | | | ChE 5303, 5304 | 3 | 3 |
| Chemical Microscopy | Chem 530 | - | 3 | ChE 5851 | 3 | (3) | ChE 5851 | 3 | (3) |
| or Special Methods of Chemical Analysis | | | | Chem 240 | (3) | 3 | | | |
| or Cost Accounting | | | | | | | Engr 3253 | (3) | 3 |
| | | **17** | **17** | | **20** | **20** | | **19** | **19** |

84

| | | 1937–38 | | | 1947–48 | | | 1957–58 | |
|---|---|---|---|---|---|---|---|---|---|
| *Terms 7,8* | | *F* | *S* | | *F* | *S* | | *F* | *S* |
| Unit Operations of Ch.E. | ChE 705 | 3 | 3 | ChE 5303,5304 | 3 | 3 | | | |
| Chemical Engineering Lab. | ChE 710 | 2 | 2 | | | | | | |
| Unit Operations Lab. | | | | ChE 5353,5354 | 3 | 3 | ChE 5353, 5354 | 3 | 3 |
| Heat Power Engineering | ME 3P33,3P34 | 3 | 3 | ME 3535,3536 | 3 | 3 | | | |
| Mechanical Engr. Lab. | ME 3X33,3X32 | 3 | 3 | | | | | | |
| Materials Testing Lab. | | | | ME 1231 | 3 | - | | | |
| Electrical Engineering | | | | | | | EE 4931, 4932 | - | 3 |
| Advanced Inorganic Chem. | Chem 130 | 3 | 3 | | | | | | |
| Advanced Physical Chem. | Chem 420 | 3 | - | | | | | | |
| Advanced Quant. Analysis | Chem 230 | - | 3 | | | | | | |
| Special Topics in Chem. | Chem 910 | 1 | - | | | | | | |
| Chem. Eng. Thermodynamics | | | | ChE 5103,5104 | 3 | 2 | ChE 5103, 5104 | 3 | 3 |
| Library Use and Patents | | | | ChE 5711 | - | 1 | ChE 5711 | 1 | - |
| Plant Inspection | | | | ChE 5701 | - | 1 | ChE 5701 | - | 1 |
| Statistical Methods | | | | | | | ChE 5745 | 3 | - |
| Science in Western Civilization | | | | | | | Hist 165, 166 | 3 | 3 |
| Electives | | | | | | 3 | 6 | | 6 | 6 |
| | | 18 | 17 | | 18 | 19 | | 19 | 19 |

| | | 1937–38 | | | 1947–48 | | | 1957–58 | |
|---|---|---|---|---|---|---|---|---|---|
| *Terms 9,10* | | *F* | *S* | | *F* | *S* | | *F* | *S* |
| Chemical Plant Design | ChE 730 | 3 | 3 | ChE 5605,5606 | 2 | 2 | ChE 5605, 5606 | 2 | 2 |
| Electrical Engineering | EE 405,406 | 4 | 4 | EE 4951,4952 | 4 | 4 | EE 4932, 4933 | 3 | 3 |
| Machine Design | ME 3D34,3D36 | 3 | - | | | | | | |
| Mechanical Engr. Lab. | ME 3X43 | 2 | - | | | | | | |
| Industrial Organization | ME 3I41 | 2 | - | | | | | | |
| Economics | Econ 3 | - | 3 | | | | | | |
| Chemical Equipment | | | | ChE 5603,5604 | 2 | 2 | ChE 5603, 5604 | 2 | 2 |
| Chemical Engr. Computations | | | | ChE 5503,5504 | 2 | 2 | ChE 5503, 5504 | 2 | 2 |
| Senior Project | | | | ChE 5953,5954 | 3 | 3 | ChE 5953, 5954 | 2 | 2 |
| Chemical Engr. Economics | | | | ME 3253 | 3 | (3) | ChE 5746 | 3 | - |
| or Elective | | | | | (3) | 3 | | | |
| Electives | | 3 | 7 | | 4 | 4 | | 3 | 6 |
| | | 17 | 17 | | 20 | 20 | | 17 | 17 |

| | 1937–38 | 1947–48 | 1957–58 |
|---|---|---|---|
| TOTALS FOR FIVE YEARS | 176 | 190 | 176 |

| | 1967–68 | | | 1977–78 | | | 1987–88 | | |
|---|---|---|---|---|---|---|---|---|---|
| *Terms 1,2* | | *F* | *S* | | *F* | *S* | | *F* | *S* |
| General Chemistry | Chem 115,116 | 4 | 4 | Chem 207,208 | 4 | 4 | Chem 207, 208 | 4 | 4 |
| Calculus for Engineers | Math 191,192 | 4 | 4 | Math 191,192 | 4 | 4 | Math 191, 192 | 4 | 4 |
| Introductory Physics | Phys 121,122 | 3 | 3 | Phys 112 | - | 4 | Phys 112 | - | 4 |
| Introduction to Engr. | Engr 104 | 3 | (3) | Engr 105,106 | 3 | 3 | Engr 110 | 3 | - |
| or Engr. Graphics | Engr 103 | (3) | 3 | | | | | | |
| or Approved Elective | | | | | | | | (3) | - |
| Liberal elective | | 3 | 3 | | | | | | |
| Nat'l or Soc. Sci. elective | | | | | 3 | - | | | |
| Elective | | | | | | | | - | 3 |
| Freshman seminar | | | | | 3 | 3 | | 3 | 3 |
| Computer Programming | | | | | | | Engr 100 | 4 | - |
| | | 17 | 17 | | 17 | 18 | | 18 | 18 |
| | | | | | | | | | |
| *Terms 3,4* | | *F* | *S* | | *F* | *S* | | *F* | *S* |
| Physical Chemistry | Chem 285,286 | 5 | 5 | Chem 287,288 | 3 | 3 | Chem 287, 288, | | |
| Physical Chemistry Lab. | | | | Chem 289,290 | 2 | 2 | 289, 290 | 5 | 5 |
| Engineering Mathematics | Math 293,294 | 4 | 3 | Math 293,294 | 4 | 3 | Math 293, 294 | 4 | 4 |
| Physics | Phys 233,234 | 3 | 3 | Phys 213,214 | 4 | 4 | Phys 213, 214 | 4 | 4 |
| Physics Laboratory | Phys 235,236 | 1 | 1 | | | | | | |
| Mass & Energy Balances | ChE 5101 | 3 | - | ChE 110 | 3 | - | ChE 219 | 3 | - |
| Equil. and Staged Ops. | ChE 5102 | - | 3 | | | | | | |
| Engr. Science elective | | 3 | 3 | | - | 3 | | | |
| Engr. Distribution course | | | | | | | | - | 3 |
| Liberal Studies elective | | | | | 3 | 3 | | | |
| Humanities or Social Science elective | | | | | | | | 3 | 3 |
| | | 19 | 18 | | 19 | 18 | | 19 | 19 |
| | | | | | | | | | |
| *Terms 5,6* | | *F* | *S* | | *F* | *S* | | *F* | *S* |
| Organic Chemistry | Chem 357,358 | 5 | 5 | Chem 357,358 | 3 | 3 | Chem 357, 358 | 3 | 3 |
| Organic Chemistry Lab. | | | | Chem 251 | 2 | - | Chem 251 | 2 | - |
| Anal. of Stage Processes | 5303 | 3 | - | | | | | | |
| Equil. and Staged Ops. | | | | ChE 311 | 3 | - | | | |
| Rate Processes | ChE 5304 | - | 3 | ChE 430 | 3 | - | | | |
| Fluid Mechanics | | | | | | | ChE 323 | 3 | - |
| Heat and Mass Transfer | | | | | | | ChE 324 | - | 3 |
| Separation Processes | | | | ChE 431 | - | 3 | ChE 332 | - | 4 |
| Chemical Processes | ChE 5203 | - | 4 | | | | | | |
| Chemical Microscopy | ChE 5851 | 3 | (3) | | | | | | |
| Materials | | | | ChE 321 | - | 4 | | | |
| Thermodynamics | | | | | | | ChE 313 | 4 | - |
| Reaction Kinetics | | | | | | | | | |
| and Reactor Design | | | | | | | ChE 390 | - | 3 |
| Mechanics | Engr 211,212 | 4 | 4 | | | | | | |
| Core science | | | | | | | | | |
| or technical elective | | | | | 3 | - | | | |
| Technical or free elective | | | | | - | 3 | | | |
| Liberal electives | | 3 | 3 | | 3 | 3 | | | |
| Humanities or Social Science elective | | | | | | | | 3 | 3 |
| | | 18 | 19 | | 17 | 16 | | 15 | 16 |

86

| Terms 7, 8 | | 1967–68 F | S | 1977–78 F | S | 1987–88 F | S |
|---|---|---|---|---|---|---|---|
| Thermodynamics | ChE 5103 | 3 | - | ChE 312 | 3 - | | |
| Reaction Kinetics | ChE 5106 | - | 3 | ChE 410 | - 3 | | |
| Unit Operations Lab. | ChE 5353 | 3 | - | Che 432 | 3 - | | |
| Chem. Eng. Lab. | | | | | | ChE 432 | 4 - |
| Project Laboratory | ChE 5354 | - | 3 | | | | |
| Polymeric Materials | ChE 5742 | 3 | - | | | | |
| Materials | ChE 5256 | - | 4 | | | | |
| Chemical Processes | | | | ChE 461 | 3 - | | |
| Chemical Process Synthesis | | | | ChE 462 | - 4 | | |
| Chemical process or system elective* | | | | | | | 3* - |
| Chemical Process Design | | | | | | ChE 462 | - 4 |
| Chemical Process Control | | | | | | ChE 472 | - 3 |
| Nonresident Lectures | | | | ChE 101 | - 0 | | |
| Electrical Engineering | EE 341, 342 | 3 | 3 | | | | |
| Liberal electives | | 3 | 3 | | 3 3 | | |
| Humanities or Social Science elective | | | | | | | 3 3 |
| Electives | | 3 | 3 | | 3 6 | | 6 6 |
| | | 18 | 19 | | 15 16 | | 16 16 |

TOTAL FOR FOUR YEARS     144     136     137

*One course from ChE 563, 566, or 643

## M.Eng.(Chemical) Curricula

| Terms 9, 10 | | 1967–68 F | S | 1977–78 F | S | 1987–88 F | S |
|---|---|---|---|---|---|---|---|
| Process Design and Economics | ChE 5621 | 6 | - | | | | |
| Process Equipment Design and Economics | | | | ChE 563 | 3 - | | |
| Process and Plant Design | ChE 5622 | - | 6 | | | | |
| Design Project | | | | ChE 565 | - 3 | ChE 565 | - 3 |
| Design of Chem. Reactors | | | | ChE 564 | - 3 | | |
| Process Control | ChE 5717 | - | 3 | ChE 671 | - 3 | | |
| Chem. Eng. fundamentals* | | | | | | | 3* 3* |
| Applied Chem. Eng. science** | | | | | | | 3** 3** |
| Higher Calculus | Math 315 | 4 | - | | | | |
| Numerical Methods in Chemical Engineering | | | | ChE 651 | 3 - | | |
| Technical electives | | 6 | 6 | | 9 6 | | 9 6 |
| | | 16 | 15 | | 15 15 | | 15 15 |

TOTAL FOR FIVE YEARS     175     166     167

*Two courses from ChE 711, 713, 731, 751
**Two courses from ChE 563, 564, 566

Appendix D

# Metallurgical Engineering Curriculum

1957–58

### Terms 1,2

| | | F | S |
|---|---|---|---|
| Inorganic Chemistry | Chem 113,114 | 4 | 4 |
| Anal. Geom. & Calculus | Math 161,162 | 3 | 3 |
| Experimental Physics | Phys 115,116 | 3 | 3 |
| English | Engl 111,112 | 3 | 3 |
| Drawing & Desc. Geom. | Engr 3117,3118 | 2 | 2 |
| Fund. of Machine Tools | Engr 3403 | 1 | - |
| Orientation | Engr 5000 | 0 | 0 |
| | | 16 | 15 |

### Terms 3,4

| | | F | S |
|---|---|---|---|
| Quantitative Analysis | Chem 224 | 4 | - |
| Organic Chemistry | Chem 301 | 2 | - |
| Analytics and Calcululus | Math 163 | 3 | - |
| Applied Mathematics | Engr 1156 | - | 3 |
| Physics | Phys 117, 118 | 3 | 3 |
| Metal. Raw Materials | Geol 712 | 3 | - |
| Intro. Metallurgy | MetE 6111 | 2 | - |
| Metal. Calculations | MetE 6501 | - | 2 |
| Mechanics | Engr 1151 | - | 3 |
| Cost Accounting | Engr 3253 | - | 3 |
| Public Speaking | P.S. 101 | - | 3 |
| | | 17 | 17 |

### Terms 5,6

| | | F | S |
|---|---|---|---|
| Physical Chemistry | Chem 403,404 | 3 | 3 |
| Physical Chemistry Lab. | Chem 411,412 | 2 | 2 |
| Chemical Microscopy | ChE 5851 | 3 | - |
| Materials of Construction | ChE 6255,6256 | 3 | 3 |
| Smelting and Refining | MetE 6203 | 3 | 3 |
| Intro. Metallography | MetE 6811 | - | 3 |
| Mechanics | Engr 1152 | 3 | - |
| Strength of Materials | Engr 1153 | - | 3 |
| | | 17 | 17 |

### Terms 7,8

| | | F | S |
|---|---|---|---|
| Thermodynamics | MetE 6103 | 3 | - |
| Unit Processes of Metallurgy | MetE 6253, 6254 | 3 | 3 |
| Physical Metallurgy | MetE 6311, 6312 | 2 | 2 |
| Phys. Metallurgy Lab. | MetE 6351 | 3 | - |
| Metallurgy of Casting, Working and Welding | MetE 6114 | - | 3 |
| Adv. Process Metallurgy | MetE 6221 | - | 3 |
| Plant Inspections | MetE 6701 | - | 1 |
| Library Use and Patents | ChE 5711 | 1 | - |
| Electrical Engineering | EE 4931 | - | 3 |
| Electives | | 6 | 3 |
| | | 18 | 18 |

### Terms 9,10

| | | F | S |
|---|---|---|---|
| Statistical Methods in Process Industries | ChE 5745 | 3 | - |
| Senior Project | MetE 6953,6954 | 2 | 2 |
| Metallurgical Design | MetE 6602 | - | 3 |
| Electrical Engineering | EE 4932,4933 | 3 | 3 |
| Science in Western Civilization | Hist 165,166 | 3 | 3 |
| Electives | | 8 | 6 |
| | | 19 | 17 |

TOTAL FOR FIVE YEARS     171

Appendix E

# Chemical Engineering Degrees Awarded

Based on calendar year (January-December):

| Year | Chem.E. | B.Ch.E. | B.S.Ch.E. | B.S. | M.Ch.E. | M.Eng.(Chem.) | M.S. | Ph.D. |
|---|---|---|---|---|---|---|---|---|
| 1932 | | | | | | | | 2 |
| 1933 | 3 | | | | | | | |
| 1934 | 5 | | | | | | | 1 |
| 1935 | 2 | | | | | | | 2 |
| 1936 | 7 | | | | 2 | | | |
| 1937 | 7 | | | | | | 1 | |
| 1938 | 14 | | | | | | | |
| 1939 | 8 | | | | 2 | | | |
| 1940 | 13 | | | | 1 | | | |
| 1941 | 16 | | | | | | | 1 |
| 1942 | 18 | | | | 3 | | | 1 |
| 1943 | | 27 | 11 | | | | | 2 |
| 1944 | | 48 | 36 | | | | 6 | 4 |
| 1945 | | 5 | 17 | | | | 8 | 3 |
| 1946 | | 5 | 1 | | | | 14 | 2 |
| 1947 | | 55 | | | 6 | | 1 | |
| 1948 | | 45 | | | 3 | | | 2 |
| 1949 | | 68 | | | 2 | | 1 | 2 |
| 1950 | | 74 | | | 2 | | | 4 |
| 1951 | | 45 | | | 3 | | 1 | 6 |
| 1952 | | 35 | | | 2 | | | 1 |
| 1953 | | 25 | | | | | 3 | 1 |
| 1954 | | 36 | | | | | 1 | 3 |
| 1955 | | 22 | | | | | 1 | 3 |
| 1956 | | 29 | | | | | 1 | 1 |
| 1957 | | 34 | | | | | 2 | 1 |
| 1958 | | 45 | | | 2 | | 2 | 6 |
| 1959 | | 37 | | | 1 | | 3 | 1 |
| 1960 | | 39 | | | 1 | | 4 | 3 |
| 1961 | | 48 | | | 5 | | 3 | 5 |
| 1962 | | 38 | | | 3 | | 2 | 1 |
| 1963 | | 34 | | | 5 | | 2 | 5 |
| 1964 | | 46 | | | 1 | | 1 | 8 |
| 1965 | | 39 | | 36 | 1 | | 6 | 5 |
| 1966 | | 9 | | 41 | | 46 | 6 | 5 |
| 1967 | | 4 | | 35 | | 39 | 3 | 5 |
| 1968 | | | | 35 | | 20 | 9 | 4 |
| 1969 | | 2 | | 38 | | 16 | 8 | 9 |
| 1970 | | | | 39 | | 5 | 5 | 3 |
| 1971 | | | | 39 | | 12 | 10 | 3 |
| 1972 | | | | 43 | | 14 | 7 | 5 |
| 1973 | | | | 44 | | 10 | 9 | 5 |
| 1974 | | | | 55 | | 15 | 10 | 4 |
| 1975 | | | | 34 | | 10 | 4 | 3 |
| 1976 | | | | 38 | | 9 | 8 | 1 |
| 1977 (January–June only) | | | | 61 | | 5 | 5 | 3 |

Based on academic year (July–June):

| Year | Chem.E. | B.Ch.E. | B.S.Ch.E. | B.S. | M.Ch.E. | M.Eng.(Chem.) | M.S. | Ph.D. |
|---|---|---|---|---|---|---|---|---|
| 1977–78 | | | | 67 | | 7 | 14 | 1 |
| 1978–79 | | | | 72 | | 3 | 9 | 3 |
| 1979–80 | | | | 66 | | 7 | 5 | 3 |
| 1980–81 | | | | 62 | | 9 | 4 | 1 |
| 1981–82 | | | | 72 | | 11 | 9 | 4 |
| 1982–83 | | | | 78 | | 11 | 12 | 6 |
| 1983–84 | | | | 68 | | 10 | 14 | 5 |
| 1984–85 | | | | 65 | | 10 | 7 | 5 |
| 1985–86 | | | | 34 | | 12 | 4 | 4 |
| 1986–87 | | | | 33 | | 13 | 3 | 10 |

GRAND TOTAL: 2942 degrees awarded

Appendix F

# Chemical Engineering Advisory Council

*Previous Council Members*

| Name | CU Degs. | Title | Affiliation | Period |
|---|---|---|---|---|
| John D. Baldeschwieler | BChE'56 | Professor of Chemistry | CalTech | 1981–83 |
| David S. Barmby | | Director, Advanced Tech. | Sun Company | 1984–87 |
| Samuel W. Bodman | BChE'61 | President | Fidelity Mgmt. & Res. Corp. | 1981–82 |
| James R. Donnalley, Jr. | PhD'44 | Vice-President | General Electric | 1981–85 |
| James Gillin, Jr. | BChE'47, PhD'51 | Pres., MSD-AGVET Div. | Merck and Co. | 1981–87 |
| Marjorie Leigh Hart | BChE'51 | Senior Advisor | Exxon Corp. | 1981–86 |
| Wayne C. Jaeschke | BChE'61 | Corporate Vice-Pres. | Stauffer Chemical | 1981–86 |
| Kenneth C. Kennard | | Gen. Mgr. & Vice-Pres. | Eastman Kodak | 1984–87 |
| Anthony J. Silvestri | | Mgr., Products Research | Mobil R & D | 1981–86 |
| Stanford H. Taylor | BChE'51 | Executive Vice-Pres. | AeroVironment, Inc. | 1984–87 |
| John A. Weaver | BChE'59 | Business Director | Rohm and Haas | 1981–86 |

*Present Council*

| Name | CU Degs. | Title | Affiliation | Period |
|---|---|---|---|---|
| Andreas Acrivos | | Professor of Chem. Engr. | Stanford University | 1986– |
| Leonard A. Barnstone | PhD'65 | Sr. Engr. Assoc. | Exxon Res. & Engr. | 1984– |
| H. Ted Davis | | Head, Chemical Engr. | University of Minnesota | 1987– |
| Samuel V. Fleming | BChE'63 | Senior Vice-Pres. | Arthur D. Little, Inc. | 1986– |
| Edward J. Holden | BME'65, MEng(Ind)'68 | Mgr. Tech. Services | IBM | 1988– |
| David S. Laity | | Vice-President | Chevron Res. Co. | 1981– |
| L. Gary Leal | | Prof. of Chemical Engr. | CalTech | 1981– |
| Bryce I. MacDonald | BChE'45 | Mgr. Remed. Projects | General Electric Co. | 1983– |
| W. Thomas Mitchell | | Genl. Mgr., Biochem. | Eastman Kodak | 1987– |
| George W. Roberts | BChE'61 | General Manager | Air Products & Chemicals | 1981– |
| S. C. Roberts | | Business Manager | Shell Oil Co. | 1987– |
| John E. Schmutz | BChE'55 | Sr. V.P. & Genl. Counsel | Du Pont | 1981– |
| William R. Schowalter | | Chairman, Chem. Engr. | Princeton University | 1984– |
| T. K. Smith | BChE'62 | Vice-President | Dow Chemical | 1987– |
| Larry F. Thompson | | Head, Lithographic Matls. | AT&T Bell Lab | 1988– |
| Heinn F. Tomfohrde III | BChE'56 | President | GAF Chemicals | 1988– |
| Vern W. Weekman | | Mgr., Central Res. Lab. | Mobil R&D Corp. | 1986– |
| James Wei | | Head, Chem. Engr. | MIT | 1986– |

Appendix G

# Books by Faculty Members

| | | | |
|---|---|---|---|
| Rhodes, F. H. | *Technical Report Writing* | McGraw-Hill | 1941 |
| | *Elements of Patent Law* | Cornell U. Press | 1949 |
| Winding, C. C. | *Plastics: Theory and Practice* (with R. L. Hasche) | McGraw-Hill | 1947 |
| | *Polymeric Materials* (with G. D. Hiatt) | McGraw-Hill | 1961 |
| Mason, C. W. | *Handbook of Chemical Microscopy* (with E. M. Chamot) | Wiley | 1938 |
| | Vol. I only | Wiley | 1958 |
| | | Wiley | 1983 |
| | *Introductory Physical Metallurgy* | American Society for Metals | 1947 |
| Smith, J. C. | *Unit Operations of Chemical Engineering* (with W. L. McCabe) | McGraw-Hill | 1956, 1967, 1976 |
| | (with W. L. McCabe and P. Harriott) | McGraw-Hill | 1985 |
| Harriott, P. | *Process Control* | McGraw-Hill | 1964 |
| | *Unit Operations of Chemical Engineering* (with W.L.McCabe and J.C.Smith) | McGraw-Hill | 1985 |
| Bischoff, K. B. | *Process Analysis and Simulation* | Wiley | 1968 |
| Rodriguez, F. | *Principles of Polymer Systems* | McGraw-Hill | 1970 |
| | | Hemisphere/ Harper & Row | 1982 |
| Gubbins, K. E. | *Theory of Molecular Fluids. I: Fundamentals* | Oxford U. Press | 1984 |

Appendix H

# **Songs and Poems**

## SONG OF THE (ChE) CLASSES

Oh, we are the Freshmen; we're just starting out;
We have no idea what Chem E is about.
We drink all the time and we party till three;
We hardly can wait till we get in Chem E.

*Refrain:*
*For it's one, two, three, four; we all fall in line;*
*To the tune of our profs we must always keep time,*
*And it's work like a Turk, till your eyes ache like hell*
*In this grand institution, this school of Cornell.*

Oh, we are the Sophomores with slipsticks in hand;
We hate our electives but P. Chem's just grand.
But Uncle Ray's course causes us too much strife—
He only gives tests that are made by his wife.

*Refrain*

Oh, we are the Juniors, a-smoking our pot;
Our minds are expanding and starting to rot.
We work on Organic twelve hours a day
And study fine fashions avec J.L.A.

*Refrain*

Oh, we are the Seniors, just trying to pass;
We never get up for a ten o'clock class.
We party all day, drinking whiskey and wine;
Maybe someday we'll finish our process design.

*Refrain*

Oh, we're the professors with nothing to fear,
No matter where you go, we'll always be here.
We told our director we'd sing nothing dirty,
But you really can't trust us: We're all over thirty!

*Refrain*

Randy Johnson '76

## SONG
*(Tune: Song of the Classes)*

Oh, it's C.W.Mason and Fred Hoffman Rhodes,
Chasing behind us with sharp-pointed goads.
For five aching years, we never could tell
What day they would bust us right out of Cornell.

Anon, ca. 1954

## HANGOVER'S SONG
*(Tune: Song of the Classes)*

Oh, we are the hangovers, and this is our verse.
The freshmen are rotten, the sophomores are worse.
We don't give a damn for the whole junior class;
And as for the seniors, they can all kiss my ... Oh,

It's one, two, three four, etc.

Anon, ca. 1938

## PROFESSORS' SONG
*(Tune: Song of the Classes)*

We are professors who lecture in class
And tutor our students so most of them pass,
For without their attendance and tuition as well,
There would be no jobs for us here at Cornell.

Peter Harriott, 1986

CORNELL CHEMICAL ENGINEERS' MARCHING
AND DRINKING SONG

*Verse:*
Chemical engineers are we,
    an exclusive camaraderie;
Winding-Rhodes we followed
    in days gone by,
We salute
    with our glasses raised on high!

*Chorus:*
*Cornell! Make it loud and clear,*
    *we want them all to hear you shout.*
*Cornell! Make the gorges ring*
    *and echo back a mighty cheer.*
*Cornell! Here's to Olin Hall;*
    *we'll toast that place we know so well.*
*So shout! Break their ears!*
    *We're the chemical engineers*
*Who can claim the glorious name*
    *Cornell!*

F. Rodriguez, 1963


Chem E ALMA MATER

Far above the Chem E office
With their bits of steel,
Students plead with their professor,
"No, no, not the wheel...."

*Refrain:*
*Punch the buttons, push your pencils, file it 1, 2, 3.*
*Life is like one big all-nighter, when you go to Chem E.*

6 a.m., your lab is finished.
Graphs look mighty sleek.
But your prof says, "I can't read this."
Rewrite by next week.

*Refrain*

Springtime comes, it's graduation.
Jobs fall right in line.
But you can't get your diploma
Till you pass design.

*Refrain*

Randy Johnson '76


CORNELL CHEM E EVENING SONG

When the sun fades far away,
And the building's shut up tight,
Scheele works with his sieve tray
Making whiskey through the night.
The lights flick on; two profs burst in.
They say, "Why George, don't act so strange."
It's R.V.B. and Stevenson
Come to work the heat exchange.

Jim says, "I sure need a drink.
"Grab those beakers—lend a hand.
"Throw the bottoms in the sink.
"Ah, this back-mixed stuff's so grand."
Then Prof. Bischoff ambles by,
Sticks of chalk behind each ear:
"Sacré bleu, it's booze," he cries,
"This sure beats Budveiser beer."

"Quiet down!" Prof. Scheele shouts,
"You're all making too much noise."
But the din is heard throughout—
In troops Winding and the boys.
"Here's a toast!" Prof. Harriott yells,
"To this fine U.O. Lab brew.
"We're the finest of Cornell's
100-proof-grade Chem E crew."

Comes the morn, the booze is all gone
And the profs have all passed out.
The director makes his way
To the lab and looks about:
Bodies strewn about the hall
Steeped in that God-awful smell;
They sure must have had a ball.
Hail to thee, all hail Cornell.

Randy Johnson '76

## THERE'S A HOLE IN OLD OLIN

There's a hole in Old Olin, where once there were
    trees,
And up goes Old Olin, to cut off the breeze;
    Of brick and kaolin
    They're building Old Olin,
    They're building Old Olin,
To house the Chem E's.

How fair is Old Olin, how lovely a cage
For Dusty's poor boys of so tender an age;
    How quick and how fleetly
    Was Olin placed neatly
    To cut off completely
The vista from Sage.

How soon and how swiftly will come the bright day
When the hole will be filled where the steam
    shovels play;
    And the rocks and the rills
    Will be three-storey stills.
    And who pays the bills?
Well, they didn't quite say.

But we love thee, Old Olin, for you are the one
Who will give to the faculty tankfuls of fun.
    So we'll roll in Old Olin,
    And bowl in Old Olin,
    And skoal in Old Olin,
    And FOLD in Old Olin,
When you are all done.

Julian C. Smith, 1941

## BUSTEE'S SONG
*(Tune: Give My Regards to Broadway)*

Give my regards to Dusty,
    Remember me to Lucy too.
Tell all the Chem. E.'s on the hill
    That I am surely through.
Tell them just how I busted,
    Way up there in 710.
Get back to work, get on the ball,
    Or I'll see you in Arts next fall.

Anon, ca. 1938

## THE REYNOLDS NUMBER SONG

Sir Osborne Reynolds was a man or yore
Who liked to play with symbols that you might
    think a bore.
But he figured out a problem and acquired some
    fame,
And now the Reynolds number bears his name.

*Chorus:*
*Take a d times a v and a rho by mu,*
*Put them all together with a little bit of glue.*
*Then you've got a number that will see you through,*
*And tell you what the fluid's going to do.*

Does the syrup in the pipe flow as smooth as can
    be,
Or is it all mixed up like in a cup of tea?
Enter all the numbers and press the little key,
Laminar or turbulent, the answer will be.

Now lots of other numbers may come to mind:
Prandtl, Schmidt, and Grashof, and more of that
    kind.
But when you've got a sticky problem and are
    getting in a bind,
The old Reynolds number is the first one to find.

It's really very simple, so the profs all say,
But the gol-darned dimensions keep getting in the
    way.
The old English system has had its day,
So better switch over to SI today.

Peter Harriott, 1978

## SONG OF THE YELLOW RUBBER DUCK
*(Tune: Home on the Range)*

Oh give me a home where the froth factors roam,
And the water and methanol play;
Where seldom is heard an encouraging word,
And faces are gloomy all day.

Home, home in U.O., etc.

Anon, ca. 1981

## THE MASTER'S MISTAKE

This, then, is the fearsome tale
    That bald and hoary sages tell,
Whose blackened horror long has stained
    The history of Cornell.

Long, long, ago, the story goes,
    When men were strong and lusty—
When Sibley's heating system worked,
    And Rhodes was not so dusty—

A bright young man arrived in town,
    Intent on Engineering.
"...Place every one" and "High-paid jobs"
    And such had he been hearing.

A very brilliant lad was he;
    He little knew— how could he?—
His fate was sealed when he signed up
    To be a Ch E.

For three long years he worked away
    Obtaining education;
Exuding larger quantities
    Of carmine perspiration.

The brilliant sparkle left his eye,
    His mind had lost its sureness;
(And probably he'd also lost
    A little of his pureness);

And in the fourth year down he went,
    As have so many men.
The little snap two-hour course
    That's known as 710

Spelled out his doom; but not because
    He failed, it's strange to say;
Not one report went down the drain—
    It was the other way.

'Twas one dark day in Baker Lab,
    That Dusty made his error
A very human slip, it's true—
    Which magnifies its terror.

For carelessly, forgetting that
    His honor was at stake,
While marking up this lad's report,
    He passed him by mistake.

He swiftly saw what he had done,
    And called to all his minions;
Described to them his fearful scheme
    That rocked them on their pinions.

The ghouls, the dwarfs, the one-eyed gnomes,
    The monsters, to a man,
Bellowed and howled and wildly cheered
    The mad unholy plan.

Then quietly they took this boy
    And tapped him on the head;
And sanded smooth his surfaces,
    And plated him with lead;

And shipped him off to MIT,
    Where still he stands today,
To show what happens when a fiend
    Finds something in his way.

*Envoi*

Remember, then, this awful tale,
    Whose horror does not fade....
The Master may forget again,
    And give a passing grade.

Julian C. Smith
*Read at Chemical
Engineers' Banquet,
November 1940*

PACKED COLUMN FATIGUE (U. O. BLUES)

*"I know what I mean, but I can't say it."*

The style book lies unopened when
The fledgling writer takes his pen,
And twisting to a new position
Tries his hand at exposition.

With all the elegance and grace
Of camels falling on their face,
The words, set down in tortured rage,
Limp and stumble across the page.

Weary the hand and weary the mind;
Eyes grow heavy and almost blind.
Dawn comes early; here's the sun
And rest at last. The damn thing's done!

"Abstract: *Results were taken on a packed column
    which was compared with published data from
    references cited in the literature.*"
                — Anon.

Maybe my abstract should say more
Than, "This lab really was a bore."
People are apt to be unshaken
On hearing the news, "Results were taken."

A word or two of what I learned
Might keep my tail from getting burned.
Perhaps a number, here and here,
Would help to make my meaning clear.

Clear? Now wait: My goal, my creed,
Is to obscure, conceal, mislead,
Keep the media unexcited.
You never know, I may be indicted.

*"The objective of this lab was to pick up the flow
    characteristics of the column."*
                — Anon., Jr.

The column, now in stately sinuous motion,
Flows gently through the laboratory air,
Arousing little interest or emotion,
But drops a characteristic here and there.

*"The irrigated pressure drops are all above the dry
    packing and both are greater than the Ergun
    equation."*
                — Ibid.

Dancing around a blazing rod of fire
We celebrate the misplaced modifier:
The dry graph dessicates the data raw;
The wet graphs ooze their statements of the Law.
The wetted runs and wetted situation
Clasp dripping hands and join the celebration,
While irrigated data gladly serve
The Ergun correlation straight-line curve.

*"The dry-packing-pressure-drop-air-velocity-
    correlation discussion appears on page 11."*
                — Op. cit.

This is the way I write (I scorn the purist's frown)
Enamored of the adjectival noun—
Multi-hyphenate rising, many-splendored, tall,
And utterly incomprehensible to all.

*             *             *

Nights and days upon the rack
To get it done— and now it's back.
What reason could there be for me
To suffer such indignity?

Write it short or write it long,
No matter what, they'll call it wrong.
Of all the words of day or night
The saddest are Revise! Rewrite!

Now abide in faith, hope, and clarity.
I need 'em all.

                Julian C. Smith, 1973

THE U.O. BLUES— PART II
or
PLEASE, PROFESSOR SMITH, WON'T YOU LEVA
US ALONE!

The cry was passed from mouth to ear,
 "Our labs are back! They're finally here!"
I plunged headlong for my mailbox,
 Totally ignoring the horrible shocks
I'd heard, from the most reliable sources.
 Not me— I've had three English courses.
A steady hand reached into the niche,
 But fingers soon began to twitch
As eyes no number grade could see,
 Just three small letters— R.I.P.

I hate reports! I canot write them
 Without mistakes *ad infinitum.*
My data points, forever sloping,
 Left me for a straight line groping.
The tower packing, I'd forgotten;
 My Leva values, ill-begotten.
My Ergun plot, it seems, I screwed
 By nineteen orders of magnitude.
Discrepancies, though, I must disclaim;
I'm *sure* my calculator's to blame.

Mere trembles increase to violent shakes
 As eyes meet more, yes, *more* mistakes.
This man must find extreme perversion
 In ruining my factors of conversion.
What's this he says? No one converts
 From psi to megahertz?
Aren't there ten feet in a liter,
 And about a yard per rotameter?
I wish I came from Tennessee,
 Where *pi* was once reduced to three.

I turn the page, in morbid terror,
 To find yet one more "minor" error.
My apparatus and procedure
 Neglected one important feature—
More than one, to be precise,
 For I left out the whole device.
Meters, headers, spray distributors
 To me, anyway, were non-contributors.
I discover, too, he really docks
 For leaving ®'s off "Intalox."

My hands, now clenched in savage rage,
 Relax. I somehow turn the page
And see, where once were words, just red.
 I swear to God, someone has bled
All over my poor composition
 Without a shred of inhibition.
Slashes, daggers, circles, and stars
 Litter the page like surgical scars
From operations that must be made
 Before I get a passing grade.

My hands are lifeless, twisted, bent
 Before the blood-stained document.
Thirty pages, in red and white
 (The black, it seems, has dropped from sight)
Reveal a newly-found humility
 Derived from countless days of futility.
I strive to check, but cannot hide
 An incessant urge for suicide.
But I can't go yet, to my great sorrow—
 The #!$(#%)! rewrite's due tomorrow!

Scott Smith, 1976

## VISCOSITY REMEMBERED

I breathe SI, life holds no joys
    Because some foreign rascal reckoned
That I must call my cp.
    A silly mPa· s.

And gone is (lb/ft)/hr.
    Is that excuse for ranting noise?
Perhaps I feel the drain of power,
    An engineer who's lost his poise.

*NOTE:  cp. = centipoise
mPa·s = millipascal-second;
(lb/ft)/hr = pounds per foot per hour*

F. Rodriguez, 1980

## OWED TO THE CONSERVATION OF MASS

There can be no more elegant creature
    Than a stoichiometry teacher.
He is one who will use all his talents
    To prove that equations must balance.

    In eleven weeks flat
    He'll convince us all that
    Whatever goes in—
    Be it ether or tin,
    Or coal dust or gin—
    It will find its way out,
    With a whoop and a shout,
Or else it will stay where it's at.

Julian C. Smith, 1977

## WHAT, THEN, IS A CHEMICAL ENGINEER?

A Chemical Engineer
is one who knows too little chemistry to be a
    chemist and too little engineering to be an
    engineer;

is someone who calculates a quantity to seven
    decimal places, then doubles it for luck;

is someone who creates polluting chemicals on the
    one hand and cleans them up on the other
    and takes credit for both;

is someone who studies the subject for four years
    at a university, with the hope of using little of
    it after five more years, and none at all after
    ten;

is someone whose office window faces a glorious
    vista of tanks and pipes filled with phosgene,
    prussic acid, and diphenyl-paraphenylene-
    alpha-magoozilide;

is someone who spends money he doesn't possess,
    on processes that haven't been tried, to make
    products that haven't been made, to fill
    markets that do not exist.

Julian C. Smith, 1977

# Index